Praise for *Mama's Last Hug*

"Game-changing. . . . For too long, emotion has been cognitive researchers' third rail. . . . But nothing could be more essential to understanding
how people and animals behave. By examining emotions in both, this
book puts these most vivid of mental experiences in evolutionary context, revealing how their richness, power and utility stretch across species and back into deep time. . . . [T]he book succeeds most brilliantly in
the stories [Frans] de Waal relates."
—Sy Montgomery, *The New York Times Book Review*

"An original thinker, [de Waal] seems to invite us to his front-row seats,
sharing the popcorn as he gets us up to speed on the plot of how life
works, through deeply affecting stories of primates and other animals,
all dramas with great lessons for our own species."
—Vicki Constantine Croke, *The Boston Globe*

"De Waal's conversational writing is at times moving, often funny and
almost always eye-opening. . . . It's hard to walk away from *Mama's Last
Hug* without a deeper understanding of our fellow animals and our own
emotions."   —Erin Wayman, *Science News*

"There is nothing sentimental about de Waal's position. . . . All the things
he has learned about animals have come from observation. He is a brilliant observer, and is often amazed by what he sees."
—Michael Bond, *New Scientist*

"I doubt that I've ever read a book as good as *Mama's Last Hug*, because it
presents in irrefutable scientific detail the very important facts that animals do have these emotions as well as the other mental features we once
attributed only to people. Not only is [the book] exceedingly important,

it's also fun to read, a real page-turner. I can't say enough good things about it except it's utterly splendid." —Elizabeth Marshall Thomas

"Before I realized Frans de Waal's connection to Mama's actual last hug, I sent the online video link to a large group of scientists saying, 'I believe it is possible to view this interaction and be changed forever.' Likewise, I believe that anyone reading this book will be changed forever. De Waal has spent so many decades watching intently and thinking deeply that he sees a planet that is deeper and more beautiful than almost anyone realizes. In these pages, you can acquire and share his beautiful, shockingly insightful view of life on earth."
—Carl Safina, author of *Beyond Worlds: What Animals Think and Feel*

"After you've read *Mama's Last Hug* it becomes obvious that animals have emotions. Learn how they resemble us in many ways."
—Temple Grandin, author of *Animals Make Us Human* and
Animals in Translation

"Frans de Waal is one of the most influential primatologists to ever walk the earth, changing the way we think of human nature by exploring its continuity with other species. He does this again in the wonderful *Mama's Last Hug*, an examination of the continuum between emotion in humans and other animals. This subject is rife with groundless speculation, ideology, and badly misplaced folk intuition, and de Waal ably navigates it with deep insight, showing the ways in which our emotional lives are shared with other primates. This is an important book, wise and accessible." —Robert Sapolsky, author of *Behave:*
The Biology of Humans at Our Best and Worst

"Another fascinating book from Frans de Waal. Once again, he makes us think long and hard about the true nature of animal emotions."
—Desmond Morris, author of *The Naked Ape*

"In *Mama's Last Hug*, Frans de Waal marshals his wealth of knowledge and experience, toggling expertly between rigorous science and captivating anecdote to explain animal behavior—humans included. While doing so, he rebukes the common conceit that we are necessarily better, or smarter, than our closest relatives."

—Jonathan Balcombe, author of *What a Fish Knows*

"De Waal is the ultimate zoological magician. His animals hold up mirrors and make you see yourself. Whether you find that terrifying or exhilarating is up to you. He is prescient, unnerving, politically explosive, and always downright entertaining. He can unmake and remake you, and you should let him."

—Charles Foster, author of *Being a Beast*

Also by Frans de Waal

Are We Smart Enough to Know How Smart Animals Are? (2016)

The Bonobo and the Atheist (2013)

The Age of Empathy (2009)

Primates and Philosophers (2006)

Our Inner Ape (2005)

My Family Album (2003)

The Ape and the Sushi Master (2001)

Bonobo: The Forgotten Ape (1997)

Good Natured (1996)

Peacemaking Among Primates (1989)

Chimpanzee Politics (1982)

MAMA'S
LAST HUG

ANIMAL EMOTIONS AND
WHAT THEY TELL US
ABOUT OURSELVES

Frans de Waal

With photographs and drawings by the author

W. W. NORTON & COMPANY
Independent Publishers Since 1923

For Catherine,
who lights my fire

Copyright © 2019 by Frans de Waal

All rights reserved
Printed in the United States of America
First published as a Norton paperback 2020

All drawings by the author. All photographs courtesy of the author, except for the photograph of one bonobo consoling another at the Lola ya Bonobo Sanctuary in the Democratic Republic of the Congo, courtesy of Zanna Clay, and the photograph of Frans de Waal holding a baby chimpanzee in 1979, courtesy of Desmond Morris.

For information about special discounts for bulk purchases, please contact W. W. Norton Special Sales at specialsales@wwnorton.com or 800-233-4830

Manufacturing by LSC Communications Harrisonburg
Book design by Patrice Sheridan
Production manager: Julia Druskin

Library of Congress Cataloging-in-Publication Data

Names: Waal, F. B. M. de (Frans B. M.), 1948– author, photographer, illustrator.
Title: Mama's last hug : animal emotions and what they tell us about ourselves / Frans de Waal; with photographs and drawings by the author.
Description: First edition. | New York : W. W. Norton & Company, [2019] | Includes bibliographical references and index.
Identifiers: LCCN 2018047218 | ISBN 9780393635065 (hardcover)
Subjects: LCSH: Emotions in animals. | Chimpanzees—Behavior.
Classification: LCC QL785.27 .W33 2019 | DDC 599.885/15—dc23
LC record available at https://lccn.loc.gov/2018047218

ISBN 978-0-393-35783-7 pbk.

W. W. Norton & Company, Inc., 500 Fifth Avenue, New York, N.Y. 10110
www.wwnorton.com

W. W. Norton & Company Ltd., 15 Carlisle Street, London W1D 3BS

1 2 3 4 5 6 7 8 9 0

CONTENTS

Prologue   1

1   MAMA'S LAST HUG   13
    An Ape Matriarch's Farewell

2   WINDOW TO THE SOUL   47
    When Primates Laugh and Smile

3   BODY TO BODY   79
    Empathy and Sympathy

4   EMOTIONS THAT MAKE US HUMAN   121
    Disgust, Shame, Guilt, and Other Discomforts

5   WILL TO POWER   171
    Politics, Murder, Warfare

6   EMOTIONAL INTELLIGENCE   203
    On Fairness and Free Will

7   SENTIENCE   239
    What Animals Feel

viii   *Contents*

8   CONCLUSION   275

Acknowledgments   279

Notes   281

Bibliography   289

Index   305

MAMA'S LAST HUG

PROLOGUE

Watching behavior comes naturally to me, so much so that I may be overdoing it. I didn't realize this until I came home one day to tell my mother about a scene on a regional bus. I must have been twelve. A boy and girl had been kissing in the gross way that I couldn't relate to but that is typical of teenagers, with open mouths moistly clamped onto each other. This by itself was nothing special, but then I noticed the girl afterward chewing gum, whereas before the kiss I had seen only the boy chewing. I was puzzled but figured it out—it was like the law of communicating vessels. When I told my mom, however, she was less than thrilled. With a troubled expression, she told me to stop paying such close attention to people, saying it was not a very nice thing to do.

Observation is now my profession. But don't expect me to notice the color of a dress or whether a man wears a hairpiece—those things don't interest me in the least. Instead, I focus on emotional expressions, body language, and social dynamics. These are so similar between humans and other primates that my skill applies equally to both, although my work mostly concerns the latter. As a student, I had an office overlooking a zoo colony of chimpanzees, and as a scientist at the Yerkes National

1

Primate Research Center, near Atlanta, Georgia, I have had a similar situation for the last twenty-five years. My chimps live outdoors at a field station and occasionally get into upheavals that cause such a ruckus that we rush to the window to take in the spectacle. What most people will see as a chaotic melee of twenty hairy beasts running about hollering and screaming is in fact a highly ordered society. We recognize every ape by face, even just by voice, and know what to expect. Without pattern recognition, observation remains unfocused and random. It would be like watching a sport that you've never played and don't know much about. You basically see nothing. This is why I can't stand American television coverage of international soccer matches: most sports narrators came late to the game and fail to grasp its fundamental strategies. They have eyes for the ball only and keep on blabbing during the most pivotal moments. This is what happens when we lack pattern recognition.

Looking beyond the central scene is key. If one male chimpanzee intimidates another by throwing rocks or charging closely past the other, you need to deliberately take your eyes off them to check the periphery, where new developments arise. I call it holistic observation: considering the wider context. That the threatened male's best buddy is asleep in a corner doesn't mean we can ignore him. As soon as he wakes up and walks toward the scene, the whole colony knows things are about to change. A female gives a loud hoot to announce the move, while mothers press their youngest offspring close.

And after the commotion has died down, you don't just turn away. You keep your eyes on the main actors—they aren't finished yet. Of the thousands of reconciliations I've witnessed, one of the first took me by surprise. Shortly after a confrontation, two male rivals walked upright, on two legs, toward each other, fully pilo-erect—meaning their hair was standing on end, making them look twice their regular size. Their eye contact looked so fierce, I expected a revival of the hostilities. But when they got close to each other, one of them suddenly turned around and presented his behind. The other responded by grooming closely around

During reconciliations after fights, chimpanzee males are eager to groom their rival's behind, which may lead to awkward 69 positions if both of them try to do so at the same time.

the anus of the first male, uttering loud lip-smacks and tooth-clacks to indicate his dedication to the task. Since the first male wanted to do the same, they ended up in an awkward 69 position, which allowed each of them to groom the other's behind at the same time. Soon thereafter they relaxed and turned around to groom each other's faces. Peace was restored.

The initial grooming location may seem odd, but remember that English (as well as many other languages) has expressions such as *brown-nosing* and *ass-licking*. I'm sure there is a good reason. Among humans, intense fear may cause vomiting and diarrhea—we say we "crap our pants" when we're frightened. That's also a common occurrence in apes, minus the pants. Bodily exits yield critical information. Long after a skirmish has ended, you may see a male chimpanzee casually stroll to the precise location in the grass where his rival had been sitting, only to bend down and take a sniff. Although vision is about as dominant a sense in chimpanzees as it is in us, smell remains critically important. In our species, too, as covert filming has demonstrated, after we shake hands with another person, especially someone of the same sex, we often scent our own hand. We lift it casually close to our face to gather a chemical whiff that informs us about the other's disposition. We do so unconsciously, as we do so many things that resemble the behavior of other primates. Nevertheless, we like to see ourselves as rational actors who know what we're doing, while we depict other species as automatons. It's really not that simple.

We are constantly in touch with our feelings, but the tricky part is that our emotions and our feelings are not the same. We tend to conflate them, but feelings are internal subjective states that, strictly speaking, are known only to those who have them. I know my own feelings, but I don't know yours, except for what you tell me about them. We communicate about our feelings by language. Emotions, on the other hand, are bodily and mental states— from anger and fear to sexual desire and affection and seeking the upper hand—that drive behavior. Triggered by certain stimuli and accompanied by behavioral changes, emotions are detectable on the outside in facial expression, skin color, vocal timbre, gestures, odor, and so on. Only when the person experiencing these changes becomes aware of them do they become feelings, which are conscious experiences. We show our emotions, but we talk about our feelings.

Take reconciliation, or a friendly reunion following a confrontation. Reconciliation is a measurable emotional interaction: to detect it, all you, as an observer, need is some patience to see what happens between former antagonists. But the feelings that accompany a reconciliation— contrition, forgiveness, relief—are knowable only to those who experience them. You may suspect that others have the same feelings as you, but you can't be sure even with respect to members of your own species. Someone may claim they have forgiven another person, for example, but can we trust this information? All too often, despite what they have told us, they bring up the affront in question on the first occasion that arises. We know our own inner states imperfectly and often mislead both ourselves and those around us. We're masters of fake happiness, suppressed fear, and misguided love. This is why I'm pleased to work with nonlinguistic creatures. I'm forced to guess their feelings, but at least they never lead me astray by what they tell me about themselves.

The study of human psychology usually relies on the use of questionnaires, which are heavy on self-reported feelings and light on actual behavior. But I favor the reverse. We need more observations of actual human social affairs. As a simple example, let me take you to a large conference in Italy, which I attended many years ago as a budding scientist. Being there to speak about how primates resolve conflicts, I hadn't expected to see a perfect human example on display. A certain scientist was acting up in a way that I had never seen before and rarely have since. It must have been the combination of him being famous and being a native English speaker. At international meetings, Americans and Brits often mistake the extraordinary privilege of being able to speak in their mother tongue for intellectual superiority. Because no one is going to disagree with them in broken English, they are rarely disabused of this notion.

There was a whole program of lectures, and after every one, our famous English-speaking scientist jumped out of his seat in the front row to help us understand the work. Just as one Italian speaker finished presenting her work, for example, and even as the applause for her lingered, this scientist rose from his seat, climbed to the podium, took the speaker's mike, and literally said, "What she actually meant to say . . ." I don't remember the topic anymore, but the Italian speaker pulled a face. It was hard to miss this man's cockiness and disrespect for her—nowadays we'd call it "mansplaining."

Most of the audience members had been listening through a translation service—in fact, their delayed linguistic connection may have helped them see through his behavior, in the same way that we're better at reading body language in a televised debate when the sound is turned off. They began to hiss and boo.

The expression of surprise on the face of our famous scientist showed how much he had misjudged the reception of his power grab. Until then, he'd thought it was going swimmingly. Flustered and perhaps humiliated, he hastily stepped down from the podium.

I kept my eyes on him and on the Italian speaker as they sat in the audience. Within fifteen minutes, he approached her and offered her his translation device, since she didn't have one. She politely accepted (perhaps without actually needing one), which counts as an implicit peace offer. I say "implicit" because there were no signs that they mentioned the previous awkward moment. Humans often signal good intentions after a confrontation (a smile, a compliment) and leave it at that. I couldn't follow what they were saying, but a third party told me that after all lectures were over, the scientist approached the speaker a second time and literally told her, "I have made a complete ass of myself." This admirable bit of self-knowledge came close to an explicit reconciliation.

Despite the ubiquity of human conflict resolution, and its fascinating unfolding at the conference, my own lecture got a mixed reception. I had only just begun my studies, and science was not yet ready for the idea that other species perform reconciliations. I don't think anyone doubted my observations—I provided lots of data and photographs to make my case—but they simply didn't know what to make of them. At the time, theories about animal conflict focused on winning and losing. Winning is good, losing is bad, and all that matters is who gets the resources. In the 1970s, science viewed animals as Hobbesian: violent, competitive, selfish, and never genuinely kind. My emphasis on peacemaking made no sense. In addition, the term sounded emotional, which was not well regarded. Some colleagues took a patronizing approach, explaining that I had fallen for a romantic notion that didn't belong in science. I was still very young, and they lectured me that everything in nature revolves around survival and reproduction, and that no organism will get very far with peacemaking. Compromise is for the weak. Even if chimps showed such behavior, they said, it's doubtful they actually needed it. And surely no other species ever did the same. I was studying a fluke.

Several decades and hundreds of studies later, we know that reconciliation is in fact common and widespread. It occurs in all social mammals, from rats and dolphins to wolves and elephants, and also in birds. The behavior serves relationship repair, so much so that if nowadays we discovered a social mammal that *didn't* reconcile after fights, we'd be surprised. We'd wonder how they kept their society together. But at the time I didn't know this and politely listened to all the free advice. It didn't change my mind, though, because for me observation trumps any theory. What animals do in real life always has priority over preconceived notions about how they ought to behave. When you are a born observer, this is what you get: an inductive approach to science.

Similarly, if you observe, as Charles Darwin famously did in *The Expression of the Emotions in Man and Animals,* that other primates employ human-like facial expressions in emotionally charged situations, you cannot get around similarities in their inner lives. They bare their teeth in a grin, they produce hoarse chuckling sounds when tickled, and they pout their lips when frustrated. This automatically becomes the starting point of your theories. You may hold whatever view you like about animal emotions or the absence thereof, but you will have to come up with a framework in which it makes sense that humans and other primates communicate their reactions and intentions via the same facial musculature. Darwin naturally did so by assuming emotional continuity between humans and other species.

Nevertheless, there is a world of difference between behavior that expresses emotions and the conscious or unconscious experience of those states. Anyone who claims to know what animals feel doesn't have science on their side. It remains conjecture. This is not necessarily bad, and I'm all for *assuming* that species related to us have related feelings, but we should not overlook the leap of faith that it asks us to take. Even when I tell you that *Mama's Last Hug* was an embrace between an old chimpanzee and an old professor a few days before her death, I cannot include her feelings in my description. The familiar behavior as well as its poignant context do suggest them, yet they remain inaccessible. This uncertainty has always vexed students of the emotions and is the reason the field is often considered murky and messy.

Science doesn't like imprecision, which is why, when it comes to animal emotions, it is often at odds with the views of the general public. Ask the man or woman in the street if animals have emotions, and they will say "of course." They know their pet dogs and cats have all sorts of emotions, and by extension they grant them to other animals as well. But ask professors at a university the same question, and many will scratch their heads, look bewildered, and ask what exactly you mean. How do you even *define* emotions? They may follow B. F. Skinner, the American behaviorist who promoted a mechanistic view of animals, by dismissing emotions as "excellent examples of the fictional causes to which we commonly attribute behavior."[1] True, it is nowadays hard to find a scientist who outright denies animal emotions, but many are uncomfortable talking about them.

Readers who feel insulted on behalf of animals by those who doubt their emotional lives should keep in mind that without the scrutiny typical of science, we'd still believe the earth is flat or that maggots spontaneously crawl out of rotting meat. Science is at its best when it questions common preconceptions. And even though I disagree with the skeptical view of animal emotions, I also feel that just affirming their existence is like saying that the sky is blue. It doesn't get us very far. We need to know more. What kind of emotions? How are they felt? What purpose do they serve? Is the fear presumably felt by a fish the same as that felt by a horse? Impressions are not enough to answer such questions. Look how we study the inner life of our own species. We bring human subjects into a room where they watch videos or play games while strapped to equipment that measures their heart rate, galvanic skin response, facial muscle contractions, and so on. We also scan their brains. For other species, we need to take the same close-up look.

I love to follow wild primates around, and over the years I've visited a great many field sites in far corners of the earth, but there's a limit to what I or anyone else can learn from this. One of the most emotional moments I ever witnessed was when wild chimpanzees high above me suddenly burst out in bloodcurdling screams and hoots. Chimps are among the noisiest animals in the world, and my heart stood still not knowing the cause of the commotion. As it turns out, they had captured a hapless monkey and were leaving little doubt about how much they prized its meat. While I watched the apes cluster around the possessor of the carcass and feast, I wondered if he shared it with them because he had more than enough to eat and didn't care or because he wanted to get rid of all those beggars, who couldn't stop whining while gingerly touching every morsel he brought to his mouth. Or perhaps, as a third possibility, his sharing was altruistic, based on how much he knew the others wanted a piece. There is no way to know for sure from watching alone. We'd need to change the hunger state of the meat owner or make it harder for the others to beg. Would he still be as generous? Only a controlled experiment would allow us to get at the motives behind his behavior.

This has worked extremely well in studies on intelligence. Today we dare speak of animal mental life only after a century of experiments on symbolic communication, mirror self-recognition, tool use, planning for the future, and adoption of another's viewpoint. These studies have blown big drafty holes in the wall that supposedly separates humans from the rest of the animal kingdom. We can expect the same to happen with respect to the emotions, but only if we adopt a systematic approach. Ideally, we'd use findings from both the lab and the field, putting them together as different pieces of the same puzzle.

Emotions may be slippery, but they are also by far the most salient aspect of our lives. They give meaning to everything. In experiments, people remember emotionally charged pictures and stories far better than neutral ones. We like to describe almost everything we have done or are about to do in emotional terms. A wedding is romantic or festive, a funeral is full of tears, and a sports match may be great fun or a disappointment depending on its outcome.

We have the same bias when it comes to animals. An Internet video of a wild capuchin monkey cracking nuts with stones will get far fewer hits than one of a buffalo herd driving lions away from a calf: the ungulates take the predators on their horns, while the calf frees itself from their claws. Both videos are impressive and interesting, but only the second one pulls at our heartstrings. We identify with the calf, hear its bleating, and are delighted by the reunion with its mother. We conveniently forget that for the lions there is nothing happy about this outcome.

That's another thing about the emotions: they make us take sides.

Not only are we keenly interested in emotions; they structure our societies to a degree that we rarely acknowledge. Why would politicians seek higher office if not for the hunger for power that marks all primates? Why would you worry about your family if not for the emotional ties that bind parents and offspring? Why did we abolish slavery and child labor if not for human decency grounded in social connectedness and empathy? To explain his opposition to slavery, Abraham Lincoln specifically mentioned the pitiful sight of chained slaves he had encountered on trips through the South. Our judicial systems channel feelings of bitterness and revenge into just punishment, and our health care systems have their roots in compassion. Hospitals (from the Latin *hospitālis*, or "hospitable") started out as religious charities run by nuns and only much later became secular institutions operated by professionals. In fact, all our most cherished institutions and accomplishments are tightly interwoven with human emotions and would not exist without them.

This realization makes me look at animal emotions in a different light, not as a topic to contemplate by itself but as capable of shedding light on our very existence, our goals and dreams, and our highly structured societies. Given my specialization, I naturally pay most attention to our fellow primates, but not because I believe their emotions are inherently more worthy of attention. Primates do express them more similarly to us, but emotions are everywhere in the animal kingdom, from fish to birds to insects and even in brainy mollusks such as the octopus.

I will only rarely refer to other species as "other animals" or "non-human animals." For simplicity's sake, I will mostly call them just "animals," even though for me, as a biologist, nothing is more self-evident than that we are part of the same kingdom. We *are* animals. Since I don't look at our own species as emotionally much different from other mammals, and in fact would be hard-pressed to pinpoint uniquely human emotions, we had better pay careful attention to the emotional background we share with our fellow travelers on this planet.

1 | MAMA'S LAST HUG

An Ape Matriarch's Farewell

One month before Mama turned fifty-nine and two months before Jan van Hooff's eightieth birthday, these two elderly hominids had an emotional reunion. Mama, emaciated and near death, was among the world's oldest zoo chimpanzees. Jan, with his white hair standing out against a bright red rain jacket, is the biology professor who supervised my dissertation long ago. The two of them had known each other for over forty years.

Curled up in a fetal position in her straw nest, Mama doesn't even look up when Jan, who has boldly entered her night cage, approaches with a few friendly grunts. Those of us who work with apes often mimic their typical sounds and gestures: soft grunts are reassuring. When Mama finally does wake up from her slumber, it takes her a second to realize what is going on. But then she expresses immense joy at seeing Jan up close and in the flesh. Her face changes into an ecstatic grin, a much more expansive one than is typical of our species. The lips of chimpanzees are incredibly flexible and

In 2016, Jan van Hooff made a final visit to Mama, an old chimpanzee matriarch, on her deathbed at Burgers Zoo. Mama had a huge
grin while embracing the professor, whom she had known for forty
years. She died a few weeks later.

can be flipped inside out, so that we see not only Mama's teeth and gums
but also the inside of her lips. Half of Mama's face is a huge smile while
she yelps—a soft, high-pitched sound for moments of high emotion. In
this case, the emotion is clearly positive, because she reaches for Jan's head
while he bends down. She gently strokes his hair, then drapes one of her
long arms around his neck to pull him closer. During this embrace, her
fingers rhythmically pat the back of his head and neck in a comforting
gesture that chimpanzees also use to quiet a whimpering infant.

This was typically Mama: she must have sensed Jan's trepidation
about invading her domain, and she was letting him know not to worry.
She was happy to see him.

Recognizing Ourselves

This encounter was an absolute first. Even though in the course of their
lives Jan and Mama had had countless grooming sessions through the
bars, no human in his right mind would walk into a cage with an adult

chimpanzee. Chimpanzees don't seem large to us, but their muscle strength far exceeds ours, and reports of horrific attacks are plentiful. Even the biggest human pro wrestler would come up short against an adult chimp. When I asked Jan if he would have done the same with any other chimp at the zoo, some of whom he had known for nearly as long, he said he's too attached to life to even think about it. Chimpanzees are so mercurial that the only humans who are safe in their presence are those who have raised them, something that didn't apply to Jan and Mama. Her being so weak now, though, changed the equation. Furthermore, she had expressed positive feelings about Jan so many times in the past that each of them had come to trust the other. This had given Jan the courage for his first and last in-person audience with the longtime queen of the colony at Burgers Zoo, in Arnhem, the Netherlands.

Over the years, I have enjoyed a similar relationship with Mama. I gave her that name precisely because of her matriarchal position. But since I now live across the Atlantic, I couldn't join in the farewell. A few months before, I'd seen Mama for the last time. Having spotted my face from a great distance among the public, she hurried to greet me despite her painful gait due to arthritis. She approached the water moat separating us with hooting and grunts, while stretching out an inviting hand. The chimpanzees live on a forested island—the largest such enclosure at any zoo—where I had watched them for an estimated ten thousand hours as a young scientist. Mama knew that later in the day, when all the apes were indoors, I'd come up to her night cage for a close-up chat.

Film crews have often exploited the predictability of our greetings. Before my arrival, they'd be standing at the ready with cameras turned on. The whole colony would be unsuspecting of what was to come, and someone would point out Mama to make sure the cameras stayed on her. Invariably, she'd be sitting there at ease, grooming or sleeping, and all of a sudden she'd either notice me or hear my voice when I called, and jump up and run forward with loud panting grunts. The crew would film it all, together with my reactions and the reactions of a few other chimps, some of whom remembered me as well. Invariably, people are impressed by Mama's memory and enthusiasm.

I have mixed feelings about these filming procedures, though. First of all, they take away from a genuine reunion between old friends. And second, I can't see what is so striking about the event. Anyone who knows chimpanzees realizes that they have excellent face recognition and long memories, so what is so special about Mama being glad to see me? Is it because we don't expect this from an exotic animal? Or is it because it indicates a bond between members of different primate species? It would be as if I visited my neighbors after a year abroad, and a whole camera team followed me around to see what would happen. After I rang the bell, the door was flung open to shouts of "There you are!"

Who would be astonished?

That we are impressed by Mama remembering me is a sign of humanity's low opinion of animal emotional and mental capacities. Students of animal intelligence in large-brained species are used to hearing loads of skepticism from fellow scientists, especially those who work on smaller-brained ones, such as rats and pigeons. These scientists often view animals as stimulus-response machines driven by instinct and simple learning, and they can't stand all this talk about thoughts, feelings, and long memories. That their views are outdated is the topic of my last book, *Are We Smart Enough to Know How Smart Animals Are?*

Jan's encounter with Mama was recorded on a cell phone.[1] When it was shown on Dutch national television, with Jan's own shaky-voiced commentary (due to the emotions of the moment), the viewers of a popular talk show were extremely moved. They posted lengthy comments to the network's website or wrote directly to Jan, declaring that they had burst into tears in front of their television sets. They were devastated, partly because of the sad context—because by then Mama's death had been announced—but also because of the very human-like way she had hugged Jan while moving her fingers in a rapid rhythm in patting his neck. The latter sight came as a shock to many people, who recognized their own behavior. For the first time, they realized that a gesture that looks quintessentially human is in fact a general primate pattern. It is often in the little things that we best see evolutionary connections. These connections, by the way, apply to 90 percent of human expressions, from the way a few measly hairs stand on our bodies when we are frightened (goose bumps) to the way men and male chimpanzees slap each other's backs in exuberance. We can see this forceful contact every spring when the chimps emerge from their building after a long winter. Finally enjoying the grass and sun, they stand around in little groups hooting, embracing, and backslapping.

At other times, we react to our obvious evolutionary links to apes with ridicule (zoo visitors often mimic the way they believe apes scratch themselves) or hilarity. We love to laugh at our fellow primates. During my lectures, I often show videos of monkeys and apes in action, and my audiences crack up over almost everything, even perfectly normal behavior. Their laughter is a sign of recognition but also of unease with the uncomfortable closeness. One of my most popular short videos, watched millions of times on the Internet, shows a capuchin monkey getting upset because the food she receives for performing a task is less attractive than her companion's food. She shakes the test chamber and slams the floor with such agitation that we have no trouble recognizing her frustration at the perceived unfairness.

Worse than hilarity is disgust, which is how people used to respond to other primates. Fortunately, this has become rare, even though people still often call primates "ugly" and are shocked when I say a male is "handsome" or a female "pretty." In former times, Westerners never saw live apes, only their bones and skins, or else gravures of them, our next of kin. When the first apes went on display, no one could believe their eyes. In 1835 a male chimpanzee arrived at London Zoo and was exhibited while clothed in a sailor suit. He was followed by a female orangutan, who was put in a dress. Queen Victoria saw the exhibit and was appalled. She couldn't stand the sight of the apes, calling them painfully and disagreeably human. Disgust at the sight of apes was in fact widespread, but how could this be, unless the apes were telling us something about ourselves that we didn't want to hear? When young Charles Darwin visited the apes at the London Zoo, he shared the queen's conclusion minus her revulsion. He felt that anyone convinced of human superiority ought to come take a look.

All these various reactions were probably triggered when Jan explained on television how special Mama was and why he had visited her on her deathbed. He himself, though, found nothing shocking, funny, or surprising about the encounter. He had just felt the need to say goodbye. It was also not an asymmetrical affair, such as when people encounter a bear, elephant, or whale, get close, and say they feel one with the animal. Humans in such situations experience an overwhelming bond and are deeply moved, but that these feelings are mutual is doubtful. Such encounters are almost like a "suicide pact," because they endanger the humans, and the animals are seriously out of luck if they get blamed for a fatal outcome.

One journalist was so enamored of a male chimpanzee in a sanctuary that when he looked into the ape's eyes, he questioned his own identity: he felt like he was staring straight into his lost evolutionary past. In his desire to show respect, however, he unintentionally ended up being condescending. Extant apes are not merely time machines to show us our own evolutionary origins! While it is true that we descend from an apelike ancestor, the ancient species that gave rise to us no longer exists. It dwelled the earth about six million years ago, and its descendants went through numerous changes and died out one by one before giving rise to the survivors around today: the chimpanzee, the bonobo, and our own species. Since these three hominids have equally long histories behind them, they are equally "evolved." So looking at an ape reveals our shared history not only to us but also to the ape looking back at us. If apes are time machines for us, then we are the same for them.

With Jan and Mama, however, none of these considerations came into play. The fact that they belonged to different species was secondary. Theirs was an encounter between two members of related species who had known each other for a long time and respected each other as individuals. We may feel mentally superior when we pet a rabbit or walk a dog, but when it comes to apes, I find this attitude impossible to maintain. Their socio-emotional lives resemble ours to such a degree that it is unclear where to draw the line.

Donald Hebb, the Canadian neuroscientist known as the father of neuropsychology, noted this blurred boundary when he studied chimpanzees at the Yerkes National Primate Research Center (now outside Atlanta, but in the 1940s it was located in Florida). He concluded that chimpanzee behavior could not be squeezed into the little definitional boxes that we put other animal behavior in, such as feeding, grooming, mating, fighting, vocalizing, gesturing, and so on. We like to write down every little thing apes do, but what is behind their behavior is hard to pinpoint. According to Hebb, we'd be much better off classifying ape behavior at the emotional level, which we grasp intuitively:

> *The objective categorization missed something that the ill-defined categories of emotion and the like did not—some order, or relationship between isolated acts that is essential to comprehension of the behavior.*[2]

Hebb was hinting at the prevailing view in biology that emotions orchestrate behavior. Taken by themselves, emotions are pretty useless: simply being fearful doesn't do an organism any good. But if a fearful state prompts the organism to flee, hide, or counterattack, it may well save its life. Emotions evolved, in short, for their capacity to induce adaptive reactions to danger, competition, mating opportunities, and so on. Emotions are action-prone. Our species shares many emotions with the other primates because we rely on approximately the same behavioral repertoire. This similarity, expressed by bodies with similar design, gives us a profound nonverbal connection with other primates. Our bodies map so perfectly onto theirs, and vice versa, that mutual understanding is close behind. This is why Jan and Mama met as equals rather than as man and beast.

You might counter that *equals* is not the right term for a free human compared with a captive ape. That is a fair comment. But Mama, born in 1957 at the Leipzig Zoo, in Germany, had no idea of life in the wild. As zoos go, she had the immense luck of joining the world's first large chimpanzee colony. In the decades since the first live specimens upset the British queen, zoos had caged the species alone or in small groups. Chimps were considered too violent to live in groups with more than a single grown male despite the fact that natural communities count many adult males, sometimes more than a dozen. As a student, Jan had spent time at an American facility in New Mexico, where NASA prepared young chimpanzees to be sent into space. There he saw firsthand the possibilities and problems of housing lots of apes together. The problems arose from the way they were fed: their caretakers dumped all the fruit and vegetables onto a single pile, which led to huge brawls that tore the social fabric apart. Around the same time Jane Goodall learned a similar lesson at her banana camp in Tanzania, which caused her to abandon the food provisioning of wild apes.

Inspired by his American experience, Jan and his brother, Antoon— the director of Burgers Zoo—decided to try social housing of chimpanzees while feeding them separately or in small family units. The result was the establishment, in the early 1970s, of a two-acre outdoor island with about twenty-five chimps, known as the Arnhem colony. Despite dire warnings from experts that it would never work, the colony thrived and over time has produced more healthy offspring than any other. Apes in the forests of Africa and Asia are currently in sharp decline, making zoo populations all the more valuable. The Arnhem colony was (and still is) a huge success and has become a model for zoos around the globe.

So even though Mama was captive, she enjoyed a long life in her own social universe, rich with birth, death, sex, power dramas, friendships, family ties, and all other aspects of primate society. She may have realized that Jan's special visit was related to her condition, but it remains unclear if she had any inkling of her own imminent demise. Do apes know about mortality? If Reo, a chimpanzee at the Primate Research Institute of Kyoto University, in Japan, is any indication, you'd suspect chimps lacked this awareness. In the prime of his life, Reo became paralyzed from the neck down as the result of a spinal inflammation. He could eat and drink but couldn't move his body. His weight continued to drop while veterinarians and students cared for him around the clock for six months. Reo recovered, but the most interesting part is how he had reacted to being bedridden. His outlook on life did not change one bit. Even when his condition looked grave to everyone around him, he teased young students by spitting water at them as he had done before his illness. He was thin as a rake but seemed unworried and never became depressed.[3]

We sometimes assume that other animals have a sense of mortality, such as a cow on her way to the slaughterhouse or a pet who disappears days before her death. Much of this is human projection, though, based on what *we* realize is coming. But do the animals realize it as well? Who says that a cat hiding in the basement during her last few days knows her end is near? Debilitated or in pain, she may simply want to be alone. Similarly, while it was obvious to us that Mama was physically on death's doorstep, we will never know if she was there in her own mind as well.

Mama was isolated in her bedroom at this time because male chimps, especially adolescents, often act like jerks by beating up easy targets. The zoo wanted to shield Mama in her condition from such abuse. Chimpanzee society is not for the meek and weak, which is precisely why the position Mama had held throughout her life was so impressive.

Mama's Central Role

Mama was exceptionally broadly built, with long, powerful arms. During charging displays, she looked very intimidating, with her hair standing on end, stamping her feet. Mama obviously didn't have the amount of muscles and hair of a male, especially not on her shoulders, which males puff up when they try to impress. But what she lacked in anatomy, she made up for with vigor. Mama was known for striking explosive blows against the large metal doors of the enclosure. She'd plant her fists wide apart on the ground and swing her entire body between her arms to deliver an ear-shattering kick with both feet against the door. This signaled that she was really worked up, and that no one should mess with her.

Even more than her physique, Mama's dominance came from her persona. She had the air of a grandmother who had seen it all and didn't take any nonsense from anybody. She demanded so much respect that the first time I looked into her eyes at face level across the water moat, I felt small. She had a habit of calmly nodding at you, to let you know she had noticed you. I had never sensed such wisdom and poise in any species other than my own. Her gaze was one of qualified friendliness: she was ready to understand and like you as long as you didn't cross her. She even had a sense of humor. Chimpanzees typically show a laugh face during frolicking games, but I have also seen it at incongruous moments, such as when a top male lets himself be chased around by an upset infant. While the colony's "big man" evades the screaming little monster, he wears a laugh expression, as though the absurdity of the situation amuses him. Mama once showed the same laugh face at an unexpected ending of a tense standoff, the way we react to a punch line.

One of my colleagues, Matthijs Schilder, was testing chimps' responses to predators. He put on a panther mask and, unknown to the chimps, hid in bushes close to the water moat around the ape island. All of a sudden, he lifted his head with the mask, so that a big cat seemed to be looking out at the chimps from the greenery. Always alert, they reacted within seconds with great alarm and fury. Giving loud, angry barks, they stormed forward to pelt the predator with sticks and stones. (The same reaction, by the way, is known among wild chimpanzees, who intensely fear leopards at night but pester them during the day.)

Matthijs had trouble avoiding the well-aimed projectiles and went into hiding at another spot.

After several confrontations he stood up straight and took off the mask to show his familiar face. The colony quickly settled down. But among all the apes, it was Mama whose expression gradually changed from anger and distress to a laugh face with half-open mouth and lips loosely covering the teeth. She held this face for a while, suggesting that she saw the joke of Matthijs's deception.[4]

Mama connected easily with everyone, both male and female, and had a support network like no other—she was a born diplomat. She also was not reluctant to enforce loyalty: she would take sides in male power struggles, choosing to support one male against another, but she would not tolerate other females expressing an alternative choice. Females who did so, who intervened in male battles on behalf of the "wrong" contender, would suddenly, later in the day, find themselves in trouble with Mama. She acted as party whip for her favorite candidate.

She made only one exception in this regard: her sidekick, Kuif, a chimpanzee also known as Gorilla, a name I used in some of my other books because of her all-black face. Kuif was more slightly built than Mama. Born at the same zoo, Kuif and Mama had a shared background that translated into a powerful alliance that continued up to Kuif's death a few years before Mama's. I never saw a single disagreement between those two females. They groomed each other frequently and always supported each other when one of them got into trouble. Kuif was the only female who could go against Mama's wishes without repercussions. She was partial to a particular male who was not Mama's favorite, but Mama ignored Kuif's support for him, as if she never noticed. In other respects, Mama and Kuif usually acted as one. A serious fight with one would automatically bring in the other, and everyone knew it, including the males, who had learned they could not handle both enraged females at once. Mama and Kuif were always there for each other and would literally scream in each other's arms after major upheavals.

Mama was not only a central figure in the colony but also took on the role of liaison with us humans. More than any other chimpanzee, she built up relationships with people she liked or perceived as important. She showed enormous respect for the zoo director, for example. Her connection with me, too, was largely on her own initiative. We often had close-up grooming sessions through the bars of her bedroom, which she shared with her friend Kuif. While my relations with Mama were relaxed, I had to be careful with Kuif, who sometimes tried to provoke me, testing me out. Chimps are always in the game of one-upmanship, always seeking the limits of your or their dominance. Kuif sometimes grabbed at me through the bars—when Mama was sitting right next to her, having her back. Your best strategy in such cases is to stay calm and act as if you barely notice; otherwise things might escalate. In later years, my relationship with Kuif changed radically for the better. After helping her raise her first surviving offspring, I became her favorite human being.

Sadly, Kuif had lost her previous babies due to insufficient lactation. Her newborns failed to thrive and shriveled away. Every time one of them died, Kuif went into a deep depression marked by rocking, self-clutching, refusing food, and issuing heart-wrenching screams. There were even hints of tears: although we are the only primates thought to shed them, Kuif would energetically rub her eyes with the backs of both fists the way children do after a good cry. Perhaps it was just an eye irritation, but curiously, it arose under precisely the circumstances when human tears flow.

Given how much Kuif suffered every time, I got the idea of helping her raise her next offspring on a bottle. But I foresaw a problem: mother apes are extraordinarily possessive, so much so that Kuif would probably not allow us to remove her baby in order to feed her. So Kuif would have to do the bottle-feeding herself. It was an audacious plan, never tried before.

Then a solution presented itself. A new baby was born in the colony to a deaf mother. In the past, this chimp had never succeeded raising her own offspring due to her inability to hear the soft baby sounds that indicate both contentment and discomfort. These vocalizations steer maternal behavior. The deaf mother might sit down on her infant, for example, without noticing its desperate whimpers. To prevent yet another failure, which had been as hard on this female as it had been on Kuif, we decided to remove her latest infant, named Roosje (or Little Rose), right after birth and give her to Kuif to adopt. We took care of the baby ourselves while we trained Kuif to handle the bottle. Then after weeks of training, we placed the wriggling infant in the straw of Kuif's bedroom.

Instead of picking up the baby, Kuif approached the bars where the caretaker and I were waiting. She kissed both of us, glancing between Roosje and us as if asking permission. Taking someone else's baby uninvited isn't well regarded among chimps. We encouraged her, waving our

I trained chimpanzee Kuif to bottle-feed her adoptive daughter Roosje. She expertly handled the milk bottle and would occasionally remove it to let Roosje breathe or burp.

arms at the infant, saying, "Go, pick her up!" Eventually she did, and from that moment on, Kuif was the most caring and protective mother one could imagine, raising Roosje as we'd hoped. She became quite talented at feeding and would even briefly remove the bottle if Roosje needed to burp, something we'd never taught her.

After this adoption, Kuif showered me with the utmost affection whenever I showed my face. She reacted to me as if I were a long-lost family member, wanting to hold both my hands, and whimpering in despair if I tried to leave. No other ape in the world did that. Our bottle training allowed Kuif to raise not only Roosje but some of her own children as well. She stayed eternally grateful for this turnaround of her life, which is why I always got such a warm welcome approaching the bedroom of Mama and Kuif.

These experiences also explain my reference here to emotions ranging from grief and affection to gratitude and awe, because this is what I felt while dealing with them. As we do with each other, and as Hebb advocated with regard to apes, we often describe behavior in terms of the emotions behind it. In my research, however, I tend to stay away from such characterizations, because in order to objectively analyze behavior, it is best to keep personal impressions out of it. One obvious way to achieve this is to document how apes behave among themselves rather than how they interact with us. Collecting the required data took most of my time, my main focus being colony politics. My project concerned the way males compete over rank, the mediating role of dominant females such as Mama, and the various ways conflicts are resolved.

This meant that I gave extensive attention to social hierarchy and the wielding of power, which topics were surprisingly controversial during the flower-power era of the 1970s. As students, my generation was anarchistic and fiercely democratic, didn't trust the authorities that ran the university (dismissing them as "mandarins," after the Imperial Chinese bureaucrats), viewed sexual jealousy as antiquated, and felt that any kind of ambition was suspect. The chimpanzee colony that I watched day in day out, on the other hand, showed all those "reactionary" tendencies in spades: power, ambition, and jealousy.

Sitting there with my shoulder-length hair, nourished by saccharine songs such as "Strawberry Fields Forever" and "Good Vibrations," I went through a truly eye-opening period. Right away, as a human being, I was struck by the similarities between us and our closest relative: every primatologist goes through this "If this is an animal, what am I?" phase. But then, like a true hippie, I had to come to grips with behavior that my generation roundly denounced but that was common in apes. Instead of letting this affect the way I looked at apes, I began to better understand my own kind.

It came down to the staple of the observer: pattern recognition. I started to notice rampant jockeying for position, coalition formation, currying of favors, and political opportunism—in my own environment. And I don't mean just among the older generation. The student movement had its own alpha males, power struggles, groupies, and jealousies. In fact, the more promiscuous we became, the more sexual jealousy reared its ugly head. My ape study gave me the right distance to analyze these patterns, which were plain as day if you looked for them. Student leaders ridiculed and isolated potential challengers and stole everybody's girlfriend while at the same time preaching the wonders of egalitarianism and tolerance. There was an enormous mismatch between what my generation wanted to be, as expressed in our passionate political oratory, and how we actually behaved. We were in total denial!

Mama, at least, was honest about power: she had it and wielded it. Initially, she even dominated three grown males who had been introduced to the colony rather late. These males were at a disadvantage when they entered the existing power structure and had trouble establishing themselves. Mama kept everyone in line, not hesitant to use brute force. She actually caused more injuries than a dominant male typically does, perhaps because a female must use harsher measures to stay on top. Later the males took over the top ranks and played their usual power games among themselves, yet Mama remained extremely influential as leader of the females. Any male who tried to rise in rank had to have her on his side, because without her he'd never get there. They all groomed Mama more than any other female, gently tickled her daughter Moniek (who acted like a spoiled princess), and never resisted when she pried food from their hands. They knew they had to stay on her good side.

Mama was an expert at mediation. Often after two male rivals fought, they were unable to reconcile, even though they seemed interested. They would hang around each other without an actual physical reunion. They would avoid eye contact. In fact, every time one of them looked up, the other would pick up a blade of grass or twig and inspect it with sudden interest. Their deadlock reminded me of two angry men in a bar.

Under these circumstances, Mama might walk up to one of them and start to groom him. Then after several minutes, she would slowly walk toward the other male. Her grooming partner would often follow, walking so closely behind her that there could be no eye contact with his rival. If he failed to follow, Mama might return to tug at his arm to *make* him follow. This showed that her mediation was intentional. After the three individuals sat together for a while, with Mama in the middle, she would simply get up and stroll away, leaving the two males to groom each other.

At other times, males who were unable to put an end to a protracted fight themselves would run to Mama. She would end up with two fully adult males, one in each arm, who wouldn't stop screaming at each other but at least had ceased fighting. Sometimes one male would reach over to grab the other, but Mama would have none of it and chased off the perpetrator. The two males usually made up by mounting, kissing, and fondling each other's genitals, after which they might discharge their tension by chasing off a low-ranking male.

One dramatic incident showed how much Mama acted as the colony's ultimate deal-maker. Nikkie, a brand-new alpha male, had gained the colony's top spot, but whenever he attempted to assert dominance, the others fiercely resisted. Being an alpha doesn't mean you can do anything you like, especially not for one as junior as Nikkie. Finally all the disgruntled apes, including Mama, chased after him, screaming loudly and barking. Nikkie, no longer so impressive, ended up sitting high in a tree, alone, panic-stricken, and screaming. Every line of escape was cut off. Every time he tried to come down, others chased him back up.

Then after about a quarter of an hour, Mama slowly climbed into the tree. She touched Nikkie and kissed him. Then she climbed down, as he followed close at heel. Now that Mama was bringing him with her, nobody resisted anymore. Nikkie, obviously still nervous, made up with his adversaries.

Alpha males rarely get to the top on their own, and Nikkie was no exception. He had reached his position with the help of an older male, Yeroen. This meant that Nikkie needed to stay on good terms with his partner. Mama seemed to understand this arrangement, because she once actively intervened when both males had a falling out. Yeroen was trying to mate with a sexually attractive female, but Nikkie immediately put up all his hair and started swaying his upper body, warning that he might interfere. Yeroen broke off his amorous advances and went screaming after Nikkie. Although Nikkie was the dominant of the two, his hands were tied—escalating a fight with the one who has made you king is never a good idea. At the same time, their common rival, the male from whom they had wrested the top position, was strutting around and asserting himself, sensing an opening. At this critical moment Mama stepped in. She went first to Nikkie and put her finger in his mouth, a common gesture of reassurance. While doing so, she impatiently nodded to Yeroen and stretched out her other hand to him. Yeroen came over and gave her a kiss on her mouth. When she withdrew from between them, Yeroen embraced Nikkie. After their reunion, both males, side by side, intimidated their common rival so as to underline their restored unity. Then everybody calmed down. Mama had put an end to chaos in the group by literally repairing the ruling coalition.

This event reflected what I call *triadic awareness,* or the understanding of relationships outside your own. Many animals obviously know whom they dominate, or whom their own family and friends are, but chimps go one step further by realizing who around them dominates whom and who is friends with whom. Individual A is aware not only of his own relationships with B and C but also of the B–C relationship. Her knowledge covers the entire triad. Similarly, Mama must have realized how much Nikkie depended on Yeroen.

Triadic awareness may even extend outside the group, as shown by Mama's reaction to the zoo director. She had little direct contact with him, yet she must have picked up on how jumpy and deferential the caretakers acted whenever he stopped by. Apes observe and learn, just as we do when we understand who is married to whom or to which family a child belongs. Experimenters have played back vocalizations and videos to explore how animals perceive their social world. From this research, we have learned that triadic awareness is not limited to apes— it has also been found in monkeys and ravens, for example. But Mama was its champion, possessing extraordinary social insight. Her central position in the colony derived from her ability to foster harmony and grasp political complexities, which allowed her to fix broken partnerships and mediate whenever tempers flared.

Alpha Female

Among human beings, alpha females are plentiful, ranging from Cleopatra to Angela Merkel. But I am struck by an everyday illustration in Bruce Springsteen's 2016 autobiography, *Born to Run.* As a young guitarist, Springsteen performed with the Castiles at many a dingy club in New Jersey, including those of teen "greasers," known for their extensive use of pomade. During performances for bouffant-haired greaser girls, the band discovered the preeminence of Kathy:

You came in, you set up your stuff, you started to play . . . and nobody moved—nobody. A very uneasy hour would pass, all eyes on Kathy. Then when you hit the right song, she'd get up and start to dance, trancelike, slowly dragging a girlfriend out in front of the band. Moments later, the floor was packed and the evening would take off. This ritual played itself out time and time again. She liked us. We found out her favorite music and played the hell out of it.[5]

Human hierarchies can be quite apparent, but we don't always recognize them as such, and academics often act as if they don't exist. I have sat through entire conferences on adolescent human behavior without ever hearing the words *power* and *sex*, even though to me they are what teen life is all about. When I bring it up, usually everyone nods and thinks it's marvelously refreshing how a primatologist looks at the world, then continue on their merry way focusing on self-esteem, body image, emotion regulation, risk-taking, and so on. Given a choice between manifest human behavior and trendy psychological constructs, the social sciences always favor the latter. Yet among teens, there is nothing more obvious than the exploration of sex, the testing of power, and the seeking of structure. Springsteen's band, for one, desperately tried to please Kathy and become friends with her, but they also had to be extremely careful. Due to the greaser guys hanging around the edges, being liked too much by a girl was hazardous, because "a murmur, a rumor, a sign of something more than friendship wouldn't be good for your health."

These are the primates I know!

Among chimpanzees, too, adolescent females arouse male competition and protection. Before reaching this age, they hardly count: they stroll around with the babies of others and play with age peers of both sexes, but no one pays them any heed. Once their first little sexual swelling develops, however, male eyes begin to follow them around. The pink balloon on their butt grows with every menstrual cycle. At the same time, they become sexually active. At first, they have trouble enticing males into sex and are successful only with their peers. But the larger their swelling, the more they intrigue the older males.

Every young female quickly appreciates how this gives her a leg up in the world. In the 1920s, Robert Yerkes, an American pioneer of primatology, conducted experiments on what he called "conjugal" relations (a misnomer, because chimpanzees lack stable bonds between the sexes). After dropping a peanut between a male and a female ape, Yerkes noted that a swollen female's privileges exceeded those of females who lacked such a bartering tool. Female chimps with a genital swelling invariably claimed the prize.[6] In the wild, males share meat after the hunt with

Female chimpanzees advertise fertility by sporting a large balloon-like swelling on their behind, which is a water-filled pink edema of the external genitalia. This conspicuous feature attracts males. When their first swelling appears, adolescent females quickly rise in status as they learn the benefits of sex appeal.

swollen females. In fact, when such females are around, males hunt more avidly because of the sexual opportunities that a successful hunt affords. A low-ranking male who captures a monkey automatically becomes a magnet for the opposite sex, which offers him a chance to mate in return for shared meat—until someone ranking above himself finds him out.

Male chimps' attraction to genital swellings may seem odd to us, given that most of us are repulsed by these brightly pink inflatables. But is it truly that different from the way men in our culture ogle breasts? The lure of frontal fleshy protuberances is in fact more puzzling, because they do not advertise fertility, whereas the swellings of chimpanzees do. While a young woman's bosom gradually expands, often aided by push-up bras and padding, she, too, becomes a magnet of male attention. She learns the power of cleavage, which gives her clout she's never enjoyed before, while also opening her up to jealousy and nasty comments from other women. This complex period in a girl's life, with its massive emotional upheavals and insecurities, reflects the same interplay between power, sex, and rivalry that adolescent female apes also go through.

A young female chimp learns the hard way that male protection is ephemeral, since it works only when males are around and attracted. A typical example of this learning curve involved Mama and Oortje, when the latter was going through her first cycles. During a fight over food, Mama slapped Oortje's back. Oortje then ran to Nikkie, the alpha male, and screamed her head off, making a din that was totally out of proportion to the minor reprimand she had received. She even pointed an accusing hand in Mama's direction.

But Oortje carried a sexual swelling, which was why Nikkie had been hanging around her the whole day. In response to her protest, he charged closely past Mama, the alpha female, with all his hair up. Mama didn't take this warning lying down. Screaming and barking, she went after Nikkie herself. There was no physical aggression between those two, however, and a few minutes later Oortje and Mama made eye contact from a distance. Mama nodded, and Oortje came over right away, and they embraced. All seemed well, especially when Mama also reconciled with Nikkie.

That same evening the chimpanzees were brought into the building, as usual, and the colony was broken up into smaller groups for the night. But after a while I heard a major scuffle. It turned out that as soon as Mama found herself alone with Oortje in an area without males, she attacked the young female in no uncertain terms. Their earlier reconciliation had been for public consumption only. It didn't mean the incident was forgotten.

The attractive periods of adolescent females, however empowering they may be, are brief and transient, and they are ignored as soon as an older female becomes swollen. This may seem counterintuitive, since we are used to human men being attracted to youthful partners, but this is not how things work among chimpanzees. In our species, attraction to youth makes evolutionary sense owing to our pair-bonding, or having stable families. Young women are both more available and more valuable because of the reproductive life that lies in front of them. Hence women's eternal quest to look young via hair-dyeing, makeup, implants, facelifts, and so on. Our ape relatives, however, know no long-term partnerships, and males are more attracted to mature mates. If several females are swollen at the same time, the males invariably lust after the older ones. This has also been reported for chimps in the wild. They practice reverse age discrimination, perhaps preferring mates with a track record who already have a couple of healthy children.

As a result, Mama became the biggest sex bomb of the group when, four years after giving birth to Moniek, she developed a swelling again. A whole crowd of males, young and old, gathered to engage in *sexual bargaining*. Instead of openly competing, which they also occasionally did, the males mostly groomed each other. They would allow one among them to mate undisturbed in exchange for a lengthy grooming session, especially one with the alpha male. Superficially, the scene looked relaxed: the object of desire looked on while all those Don Juans tended

When a female develops a genital swelling, her condition may cause intense male-male competition—curiously, expressed more in grooming than in fighting. Known as "sexual bargaining," males groom each other at a frantic rate in the female's presence. Subordinate males groom their superiors to "buy" an undisturbed mating. Here a female (*left*) patiently waits for the males to resolve their issues.

each other's hair. But there was great underlying tension. Any male who tried to approach Mama while skipping protocol was guaranteed to get into trouble.

What interests me most in these scenes is the males' obvious self-control. We tend to look at animals as emotional beings who lack the brakes that we apply. Some philosophers have even argued that what sets our species apart is our capacity to suppress our impulses, an idea tied to free will. Like so many proposals about human uniqueness, however, this one is wildly off the mark. Nothing could be less adaptive for an organism than to blindly follow its emotions. Who wants to be a loose cannon? If a cat were to immediately give in to its urge to rush after a chipmunk, instead of slowly sneaking up on it, it would fail time after time. If Mama had not waited for the right moment to attack Oortje, she would never have been able to underline her position. If males were to mate whenever they felt like it, they would constantly run into trouble with the competition. They need to appease the higher-ups, paying a price in grooming, or circumvent them by arranging a secret rendezvous behind the bushes—a common technique that requires female cooperation. All this rests on highly developed impulse control, which is part and parcel of social life. Trainers of horses, dogs, and marine mammals are intimately familiar with this capacity in their animals as it is their bread and butter.

Once at a Japanese zoo, I saw they had set up a nut-cracking station for their chimpanzees. The enclosure contained a heavy anvil stone and a smaller hammer stone attached to it with a chain. The caretakers would throw a large number of macadamia nuts into the enclosure, and all the chimps would gather hand- and footfuls, then sit down. Macadamias are one of the few types of nut that chimps can't crack with their teeth—they needed that nut-cracking station. First the alpha male cracked his nuts there, after which the alpha female did the same, and so on. The rest waited patiently for their turn. It was all extremely peaceful and orderly, and everyone managed to crack their nuts without problems. But underlying this order was violence: if one of them had tried to breach the established arrangement, chaos would have ensued. Even though this violence was mostly invisible, it structured society. And isn't human society set up like this, too? It is orderly on the surface but is backed by punishment and coercion for those who fail to obey the rules. In both humans and other animals, giving in to one's emotions without regard for the consequences is about the stupidest course of action to follow.

Mama lived in a complex society that she understood better than anyone else, including me, the human observer who had to work hard to unravel its intricacies. How she got to the top is not entirely clear, but from the many chimpanzee colonies that I have known over my career, and from observations in the wild, we know that age and personality are the main factors. Female chimpanzees rarely compete over rank, and they establish positions surprisingly quickly. Whenever zoos put chimpanzees from different sources together, the females establish ranks within seconds. One female walks up to another, the other submits by bowing, pant-grunting, or moving out of her way, and that's all there is to it. From then on, one will dominate the other. Altercations do occur, but only in a minority of cases. This is quite different from the males, who either try to intimidate each other, which may provoke a physical fight, or wait things out and fight a couple of days later. At some point, there is always a test of strength. Even once rank is established, it is never guaranteed: it always remains open to challenge. This is why the most vigorous males, often between the ages of twenty and thirty, take the top spots. They outrank the older ones, who step down the ladder one rung at a time after their career has peaked.

Females, by contrast, have an age-graded system in which being old offers an advantage. This system is naturally more stable than that of the males. One of the oldest ladies is the alpha despite the presence of younger females, who are physically stronger. They would have no trouble beating her in a fight, but their physical condition is irrelevant. Decades of research on wild chimpanzees have shown that females compete very little over status and just wait it out, a process known as queuing. If a female lives long enough, she is bound to end up in a high position. Since they live dispersed over the forest and forage mostly on their own, achieving a high rank probably does not yield enough benefits for females to take risks for it. It's not worth the trouble males go through.[7]

A female in Mama's position is often called a matriarch, but this term varies in meaning. Her status was quite different, for example, from that of an elephant matriarch: the oldest and largest cow leads a herd comprised of other females and their young, many of them family. As a chimp, in contrast, Mama navigated an infinitely more complex universe that included adult males who never stopped jockeying for position, and in which all the other females were unrelated to her. Her achieving the top rank in this milieu was not even the most remarkable aspect of her position, because she also wielded immense power. Power and rank are different things.

We measure rank by who submits to whom. Chimpanzees do it by bowing and pant-grunting. The alpha male needs only to walk around, and others will rush toward him and literally grovel in the dust while uttering panting grunts. The alpha may underline his position by moving an arm over the others, jumping over them, or simply ignoring their greeting as if he doesn't care. He is surrounded by massive deference. Mama received such rituals less often than any male would have, but all the other females would at least occasionally pay their respect to her, making her the top-ranking female. These outward signs of status reflect the *formal* hierarchy, the way the stripes on military uniforms tell us who ranks above whom.

Power is something else entirely: it is the influence an individual exerts on group processes. Like a second layer, power hides behind the formal order. To give a human example, the longtime secretary of a company boss often regulates access to her employer, whether male or female, and makes lots of small decisions by herself. Most of us recognize her immense power and are smart enough to befriend her, even though formally she stands rather low on the totem pole. In the same way, social outcomes in a chimpanzee group often depend on who is most central in the network of family ties and alliances. I have already described how Nikkie, the new alpha male, was not nearly as well respected as his senior partner, Yeroen. Nikkie held the highest rank, yet he was not very powerful, and the colony regularly rejected his rule. It was in fact Yeroen and Mama, the colony's oldest male and female, who ran the show. They had so much prestige that no one objected to their decisions. With her excellent connections and mediation skills, Mama was extraordinarily influential. All the adult males formally outranked her, yet if push came to shove, they all needed and respected her.

Her wish was the colony's wish.

Finality and Grief

When Mama's condition grew worse, with no hope of improvement, she was euthanized by a veterinarian. It was an immensely sad day, but the decision was inescapable. The zoo then did something that is rarely part of the death protocol: it offered the ape colony a chance to view and touch the corpse by leaving it in the night cage with doors open. All the comings and goings were filmed.

The videos make it obvious that the females were more interested than the males. The males mainly hit the corpse a couple of times and dragged it around. Such rough treatment may seem inappropriate, but we had seen it before: probably it is an attempt to rouse the dead. How to be sure that an individual is truly dead unless his or her responses have been thoroughly tested? Even in hospital emergency rooms, a person is declared dead only after resuscitation efforts have failed. The female chimpanzees did something similar, albeit a bit more subtle: they lifted up an arm or a foot and watched it drop down, or they looked inside the corpse's mouth, perhaps verifying the absence of breathing. When one of the females pulled at the body to move it, though, she got an earful from Geisha, Mama's adoptive daughter. Unlike the others, Geisha never took any break to eat or socialize and stayed with the corpse all the time. She acted like people do at a wake. A wake was originally a period during which mourners kept vigil over a deceased person at home. Probably humans originated wakes in the hope that their loved one would come back to life, or else to make absolutely sure that he or she was dead prior to burial.

Geisha is the daughter of Kuif. Mama had taken her under her wing after her mother's death. This was logical, given how tight Mama was with Kuif. Now, after Mama's death, it was Geisha who spent the most time with the corpse, even more than Mama's biological daughter and granddaughter. All the females visited in total silence, an unusual state for chimps. They nuzzled and inspected the corpse in various ways, or spent time grooming it.

They also brought a blanket from elsewhere, leaving it near Mama's body. This was harder to interpret, but reminded me of another chimpanzee death.

One day at the Yerkes Field Station, a popular former alpha male, Amos, was panting at a rate of sixty breaths per minute. Sweat poured from his face. We hadn't recognized his poor condition earlier, because like most males, he had hidden it for as long as he could. Males avoid showing vulnerabilities. Only after his death, a few days later, did we learn of Amos's hugely enlarged liver and multiple cancers. Since he refused to go outdoors with the rest, we kept him apart and cracked open a door to his night cage. A female friend, Daisy, regularly came to visit. She reached through the crack to tenderly groom the soft spot behind his ears. At some point, she fetched wood wool and started pushing large amounts of it toward Amos. Chimps like to build nests with such material. Daisy reached in several times to stuff the wool between Amos's back and the wall against which he was leaning, as if she realized he was in pain and would be better off leaning against something soft. It looked like the way we arrange pillows behind a patient in the hospital.

So even though we don't know how that blanket ended up near Mama's remains, we can't rule out that someone was trying to make her feel comfortable, perhaps in reaction to her icy state. The study of how apes and other animals respond to the death of others falls under *thanatology,* after the Greek god of nonviolent death, Thanatos. Grieving after death is hard to define, but in her 2013 book *How Animals Grieve,* Barbara King, an American anthropologist, proposes that a minimum requirement is that individuals who were close to the deceased markedly alter their behavior, such as by eating less, becoming listless, or guarding the spot where the dead was last seen.[8] If the deceased is an offspring, a mother may keep the smelly corpse until it falls apart, as has been seen many times. A chimp mother in a West African forest carried her dead infant for no fewer than twenty-seven days. This reaction is natural enough in the primates, who transport young on their belly or back, but it has also been observed in dolphins. A mother dolphin may keep her dead calf's body afloat for days.[9]

Individuals who lack a bond to the deceased have no reason to be affected by their death. Many pets, for example, hardly react at all when one in the same household has died. Grieving requires attachment. The stronger the bond, the stronger the reaction to its severance. This applies to all mammals and birds, including corvids (members of the crow family). When the mate of one of my jackdaws disappeared for unknown reasons, he called for days on end while scanning the sky. When she failed to return, he gave up and died a few days later. Then it was my turn to mourn the loss of two birds who had given me so much affection and pleasure and taught me that the emotional lives of birds are on a par with those of mammals.

Konrad Lorenz, the eminent Austrian ethologist, idealized the life-long pair-bonds of his geese. When one of his students pointed out that she'd noticed a few infidelities, she softened the blow by adding that this made geese "only human." Monogamy, or pair-bonding, is more typical of birds than mammals. In fact, very few primates are monogamous, and whether humans truly are is debatable. The accompanying emotions may be similar across species, though, as oxytocin is involved in all mammals. This ancient neuropeptide is released by the pituitary gland during sex, nursing, and birthing (it is routinely administered in maternity wards to induce labor), but it also serves to foster bonds among adults. People who have just fallen in love have more oxytocin in their blood than do singles, and their high concentration lasts if their relationship lasts. But oxytocin also shields pair-bonds from sexual adventures with outsiders. When married men are given this hormone in a nasal spray, they feel uncomfortable around attractive women and prefer to keep their distance.[10]

Even if we view human romantic love as special, the neural similarities with other species are striking. Larry Young, a neuroscientist colleague at Emory University, is known for his studies of two vole species. The meadow vole lives a promiscuous life, whereas the similar-looking prairie vole forms pairs in which male and female mate exclusively with each other and raise pups together. Prairie voles have far more oxytocin receptors in the reward centers of their brains than do meadow voles. As a result, they have an intensely positive association with sex, resulting in an "addiction" to the partner they had it with. Oxytocin ensures that they will bond. If these voles lose their mate, they show chemical brain changes suggesting stress and depression. They also become passive in the face of danger as if they don't care anymore if they'll live or die. So even these tiny rodents seem to know grief.[11]

The American zoologist Patricia McConnell describes how her dog Lassie reacted to the death of her best friend, Luke. These two dogs adored each other and were always together. Following Luke's death, Lassie stayed a whole day in the room with the body, lying with her head down with meltingly sad eyes and furrowed brow. The next day she regressed to stereotypical behavior from her youth, spinning around like crazy and licking and sucking her toys as if nursing. McConnell concluded that Lassie had grasped the finality of Luke's death. Otherwise why would her mood have changed so drastically?[12]

All indications are that at least some animals realize that a dead companion will never move again. When an adult male chimpanzee fell out of a tree and broke his neck, an adolescent wild female gazed at his body motionless, without interruption, for over an hour, while the males around embraced each other with nervous grins on their faces.[13] There would be no reason for such dramatic reactions if apes considered the situation of the other as being of a passing nature. Moreover, to realize its irreversibility implies an expectation about the future. We have quite a bit of scientific evidence for future orientation in primates, based on how they plan their travels or prepare tools for a task, but we rarely consider foresight connected to life and death. For obvious reasons, we lack experiments on this ability. If we call awareness of one's own demise a *sense of mortality*—for the existence of which we have no evidence outside our own species—we may call Lassie's recognition that Luke won't return a *sense of finality*. It differs from the sense of mortality in that it concerns the *other* rather than the self.

There are many similar stories of bereavement, also for cats, and of course between pets and their owners. Dogs have been known to spend years near the grave of their human companion or to return every day to the train station where they used to pick him up. I have visited statues in Edinburgh and Tokyo erected in admiration of loyal dogs named Bobby and Hachiko. The same postmortem loyalty may drive other animals. Elephants will gather the ivory or bones of a dead herd member, holding the pieces in their trunks and passing them around. Some elephants return for years to the spot where a relative died, only to touch and inspect the relics.

Quite a different hint of the sense of finality occurred one day after a Gaboon viper, a venomous snake, entered an African sanctuary. The snake aroused intense fear among the bonobos, and they all jumped back at its every move. The bonobos carefully poked at it with a stick, until finally the alpha female grabbed the snake, threw it into the air, and slammed it against the ground. After she dispatched it, no one gave any indication of expecting it to come back to life. Dead is dead. The juveniles happily dragged its lifeless body around as a toy, slinging it around their necks, even prying open its mouth to study its huge fangs. These apes must have regarded the snake's death as irreversible.

We rarely witness the actual death of a member of an ape colony, but at Burgers Zoo we did when Oortje literally dropped dead. She was one of my favorites, always a happy-go-lucky character. Oortje had been coughing, and her condition kept deteriorating despite the antibiotics we gave her. One day we saw Kuif staring into Oortje's eyes from up close. Then without any apparent reason, Kuif burst out screaming in a hysterical voice while hitting herself with spasmodic arm movements, as

frustrated chimps do. She seemed upset about something she had seen
in Oortje's eyes. Oortje herself had been silent until this point, but now
she feebly screamed back. She tried to lie down, fell off the log on which
she had been sitting, and remained motionless on the ground. A female
in another part of the building uttered screams similar to Kuif's, even
though she could not possibly have seen what had happened. Thereafter
twenty-five chimps turned completely silent. The caretaker moved all
other apes out of the way, then tried mouth-to-mouth resuscitation, to
no avail. Autopsy revealed that Oortje had suffered a massive infection
of heart and abdomen.

Primates, like those around Mama's body, react to death in much the
way humans attend to the departed: by touching, washing, anointing,
and grooming bodies before they let go. We go a step further, though, in
that we bury the dead and often give them something to take with them
on their "voyage." The ancient Egyptians filled the tombs of pharaohs
with food, wine, hunting dogs, pet baboons, and full-size sailing vessels. To make a loss bearable, and to calm our own terror of mortality,
humans often view death as a transition to a different life. We have no
evidence of this remarkable mental innovation in any other animal.

A discussion about these distinctions erupted around the 2015 discovery of *Homo naledi,* an early human relative. Its fossil remains had
been discovered deep inside a cave in South Africa. This primate had
Australopithecine-like hips but feet and teeth more typical of our own
genus. Most likely *Homo naledi* belongs on one of the dozens of side
branches of the gigantic bush that is our ancestry, but paleontologists
dislike such a proposal. They always prefer their find to sit on the tiny
branch that led to us, never mind that the chances of this being true
are minimal. This way they can claim to have found a human ancestor. But how could they try to make this point for *Homo naledi,* which
had an ape-size brain? When the scientists discovered fossil remains in a
nearly inaccessible part of the same cave system, they felt they had their
argument, because these remains must have been put there deliberately.

Only humans would be so considerate of their dead, they claimed. Their proposal was at once highly speculative and born from ignorance about how other species treat the departed.

Since chimpanzees and other primates never stay in one place for long, they have no reason to cover or bury a corpse. Were they to stay in one place, they would no doubt notice that carrion attracts scavengers, including formidable predators, such as hyenas. It would absolutely not exceed an ape's mental capacity to solve this problem by covering an odorous corpse or by moving it out of the way. Such behavior would hardly require belief in an afterlife. The same sort of practical need may have driven *Homo naledi*. At this point, we simply don't know if they moved their dead with care and concern or unceremoniously dumped them into a faraway cave to get rid of them. It may even be worse, because who says the discovered remains were dead at the moment they ended up in the cave?

It is an odd coincidence that the word *naledi* (which means "star" in Sotho-Tswana languages) is an anagram of *denial*. The discoverers of the fossil were way too eager to stress its humanity, while being in denial about how much our ancestors had in common with apes. Humans diverged from the apes about as long ago as African and Asian elephants did from each other, and they are genetically about as close or distant. Yet we freely call both of those species "elephants" while obsessing over the specific point at which our own lineage moved from being an ape to being human. We even have special words for this process, such as *hominization* and *anthropogenesis*. That there was ever such a point in time is a widespread illusion, like trying to find the precise wavelength in the light spectrum at which orange turns into red. Our desire for sharp divisions is at odds with evolution's habit of making extremely smooth transitions.

It remains unknown how widespread the sense of finality is and how much it relies on a mental projection into the future. But members of at least some species, after assuring themselves by smell, touch, and revival attempts that a loved one is gone, seem to realize that their relationship has permanently moved from present to past. How they achieve this realization is puzzling. Is it based on experience? Or do they intuitively know that death is part of life? It also reminds us that all emotions are mixed with knowledge—they wouldn't exist otherwise. When animals do something interesting, cognitive scientists sometimes say "That's just an emotion," but emotions are never simple and never separate from an evaluation of the situation. Grief in particular is far more complex than suggested by calling it an emotion. It represents the sad flip side of social bonding: loss. It may cut as deeply into the soul of some animals as it does into ours based on shared neural processes, such as the oxytocin system, and perhaps a similar awareness of life and its vulnerabilities.

For me, visits to the Burgers colony will never be the same. Mama's demise has left a giant hole for the chimpanzees as well as for Jan, myself, and her other human friends. She represented the heart of the colony. Life goes on, as it should, but individuals are unique. I don't expect to ever again encounter an ape personality as impressive and inspiring as Mama's.

2 | WINDOW TO THE SOUL

When Primates Laugh and Smile

Tickling a juvenile chimpanzee is a lot like tickling a child. The ape has the same sensitive spots: under the armpits, on the side, in the belly. He opens his mouth wide, lips relaxed, panting audibly in the same familiar *huh-huh-huh* rhythm of inhalation and exhalation as human laughter. The striking similarity makes it hard not to giggle yourself.

The ape also shows the same ambivalence as a child. He pushes your fingers away, protecting its ticklish spots while trying to escape from you, but as soon as you stop, he comes back for more, putting his belly right in front of you. At this point, you need only point to it, not even touching it, and he will throw another fit of laughter.

Laughter? Now wait a minute! A real scientist should avoid any and all anthropomorphism, which is why hard-nosed colleagues often ask us to change our terminology. Why not call the ape's reaction something neutral, like, say, vocalized panting? I've heard careful colleagues speak of "laugh-like" behavior. That way we avoid any and all confusion between the human and the animal.

The term *anthropomorphism*, which means "human form," comes from the Greek philosopher Xenophanes, who protested in the fifth century B.C. against Homer's poetry because it described the gods as though they looked human. Xenophanes mocked this assumption, saying that if horses had hands, they would "draw their gods like horses." Nowadays the term has gained a broader meaning. It is typically used to censure the attribution of human-like traits and experiences to other species. So animals don't have "sex" but engage in breeding behavior. They don't have "friends," only favorite affiliation partners. Given how partial our species is to intellectual distinctions, we apply these linguistic castrations even more vigorously in the cognitive domain. By boiling down everything animals do to instinct or simple learning, we keep human cognition on its pedestal. To think otherwise opens you up to ridicule.

To understand this resistance, we need to go back to another ancient Greek, Aristotle. The great philosopher put all living creatures on a vertical *scala naturae,* which runs from humans (closest to the gods) down toward other mammals, with birds, fish, insects, and mollusks near the bottom. Comparisons up and down this vast ladder have been a popular scientific pastime, but all we seem to have learned from them is how to measure other species by our own standards.

But how likely is it that the immense richness of nature fits on a single dimension? Isn't it rather to be expected that each animal has its own mental life, its own intelligence and emotions, adapted to its own senses and natural history? Why would the mental life of a fish and a bird be the same? Or take predators and prey. Obviously, predators have a different emotional repertoire than species that need to constantly look over their shoulder. Predators exude a cool self-confidence (except when they meet their match), whereas prey animals know fifty shades of fear. They live in terror, and they startle at every unexpected movement, sound, or smell. This is why horses bolt and dogs don't. We evolved from tree-dwelling fruit pickers—hence our frontal eyes, color vision, and grasping hands—but because of our size and special skills, we share a predator's poise. This is probably why we get along so well with our favorite pets, which are two furry carnivores.

In college, I had a black and white kitty named Plexie. About once a month, I would take Plexie on my bicycle, in a bag with her little head sticking out, for a playdate with her best friend, a short-legged puppy. The two of them had played together since they were little and kept doing so now that they were both adult. Racing up and down the stairs of a large student house, surprising each other at every turn, their obvious joy was contagious. They could go at it for hours until they'd plop down exhausted. Dogs and cats often get along well because they're both eager to chase and grab moving objects. They're also both mammals, which helps them relate to us. Other mammals recognize our emotions, and we recognize theirs. It's this empathic connection that attracts humans to domestic cats (an estimated 600 million worldwide) and dogs (500 million) rather than, say, iguanas and fish. With that human-animal connection, however, comes our tendency to project feelings and experiences onto animals, often uncritically.

We may say that our dog is "proud" of a ribbon he has won in a show or our cat is "embarrassed" that she missed a jump. We go to beach hotels to swim with dolphins, convinced that the animals must love this as much as we do. Lately, people have fallen for the claim that Koko, the late hand-signing gorilla in California, worries about climate change, or that chimpanzees have religion. As soon as I hear such suggestions, my corrugator muscles contract into a frown, and I ask for the evidence. Gratuitous anthropomorphism is distinctly unhelpful. Yes, dolphins have smiley faces, but since this is an immutable part of their visage, it fails to tell us anything about how they feel. And a dog carrying a ribbon may simply enjoy all the attention and goodies that come his way.

However, when experienced fieldworkers, who follow apes around every day in the tropical forest, tell me about the concern chimpanzees show for an injured companion, bringing her food or slowing down their walking pace, I am not averse to speculations about empathy. And when they report that adult male orangutans in the treetops vocally announce which way they'll travel the next morning, I don't mind the suggestion that they plan ahead. Given everything we know from controlled experiments in captivity, these speculations are scarcely far-fetched. But even in these cases, accusations of anthropomorphism fly.

The anthropomorphism argument is rooted in human exceptionalism. It reflects the desire to set humans apart and deny our animality. To do so remains customary in the humanities and much of the social sciences, which thrive on the notion that the human mind is somehow our own invention. I myself, however, consider the rejection of similarity between humans and other animals to be a greater problem than the assumption of it. I have dubbed this rejection *anthropodenial*. It stands in the way of a frank assessment of who we are as a species. Our brains have the same basic structure as those of other mammals: we have no new parts and employ the same old neurotransmitters. Brains are in fact so similar across the board that in order to treat human phobias, we study fear in the rat's amygdala. Dogs trained to lie still in a brain scanner show activity in the caudate nucleus when they expect a hot dog in the same way this region lights up in businessmen who are promised a monetary bonus. Instead of treating mental processes as a black box, as previous generations of scientists have done, we are now prying open the box to reveal a shared background. Modern neuroscience makes it impossible to maintain a sharp human-animal dualism.[1]

This doesn't mean that orangutan planning is of the same order as me announcing an exam in class and my students preparing for it, but deep down there is continuity between both processes. An even greater continuity applies to emotional traits. Since our understanding of emotions is partly intuitive, the continuity is hard to explain based purely on data and theory. It helps to have intimate exposure to animals, such as the kind that pet lovers enjoy every day. Hence my simple and unscientific recommendation that any academic who doubts the depth of animal emotions ought to get a dog.

Anthropomorphism is not nearly as bad as people think. With species like the great apes, it is in fact logical. Evolutionary theory almost dictates it, given that we know apes as "anthropoids," which means "human-like." We owe this term to Carl Linnaeus, the eighteenth-century Swedish biologist who based his classification on anatomy, but he could just as easily have done so on the basis of behavior. The simplest, most parsimonious view is that if two related species act similarly under similar circumstances, they must be similarly motivated. We don't hesitate to make this assumption when comparing related species like horses and zebras, or wolves and dogs, so why change the rules for humans and apes?

Fortunately, times are changing. The natural sciences have permanently blurred the human-animal divide popular in Western culture and religion. Nowadays we often start from the opposite end, assuming continuity and shifting the burden of proof to those who insist on a gap. It is up to them to convince us. Anyone who wants to make the case that a tickled ape, who almost chokes on his hoarse giggles, must be in a different state of mind from a tickled human child has his work cut out for him.

Express Yourself

Many years ago I traveled with Jan van Hooff to a workshop in the Netherlands given by Paul Ekman and his followers. The American psychologist was a celebrated guest in our country. He was not yet as famous as he would become, but he was already making waves with his research on the human face. Ekman had developed the Facial Action Coding System (FACS), which classifies facial expressions by plotting every little muscle contraction. We have, for example, a small muscle close to the eye with a Latin name that means "wrinkler of the eyebrows," and a large muscle in each cheek that lifts the mouth corners, pulling them upward, resulting in a smile. Ekman could personally demonstrate almost any configuration, having uncanny control over his own visage. He had no trouble producing the minutest movements, either symmetrically or asymmetrically, that conveyed small shifts in emotion. He could look angry, or angry masked with a broad smile, or pleased mixed with worry. You name it—his face could produce a panoply of subtle emotions on command. He would illustrate how a slight frown indicates one emotion, and a wrinkled nose another. We admired not only his facial acrobatics but also his evolutionary outlook, which at the time was exceptional in a psychologist.

I speak of "acrobatics," because obviously his work was all about movement and form. We humans are very well capable of making an annoyed-looking face without actually feeling annoyed. We have reasonable facial control. For the longest time, I thought other primates might lack it, until I studied bonobos at the San Diego Zoo. There I found myself in a situation that in retrospect is rather amusing.

I had taken on the task of documenting bonobos' entire behavioral repertoire—their calls, facial expressions, gestures, and postures—something no one had done before. But every time I observed a group of juveniles in their spacious, green enclosure, my list of facial expressions grew longer and longer. There seemed no end in sight. I had to describe the weirdest expressions, and they never matched those I'd seen before. After a while it dawned on me that the most unusual ones always occurred in nonsocial situations and never led to particular actions, such as sex or aggression, that might betray their meaning. A young bonobo would be sitting staring at nothing in particular and suddenly go through a pantomime of sucked-in cheeks, a bulging upper lip, and rapid jaw movements. Sometimes a hand would get involved—for example, pulling a lip sideways or reaching all the way around the back of the head to stick a finger into the mouth from the "wrong" side.

I concluded that the bonobos were just amusing themselves with fantasy grimaces that made no sense whatsoever. I called them "funny faces" and saw them as a sign of excellent voluntary control over the facial musculature. Could not an animal that pulls faces for fun do the same in order to manipulate others? Whatever the implications, these young apes certainly made me see the foolishness of science's obsession with classification. Once I realized what their facial acrobatics were all about, I couldn't suppress the feeling that they occasionally winked at me!

Ekman's emphasis on the facial exterior appealed to Jan and me. We study animal behavior from a biological standpoint, focusing on signals, the form they take, and their effect on others. In fact, for the longest time we were not allowed to talk about anything else! Jan had received urgent personal advice, from no one less than the Nobel Prize–winning zoologist Niko Tinbergen, to stay away from internal states in his study of primate facial expressions. Why mention emotions if you can avoid them? He'd describe the chimpanzee's laugh or play face as a "relaxed open-mouth face"; instead of a grin or smile, he'd speak of a "silent bared-teeth face"; and so on. Ekman did the same in his FACS, which was purely descriptive, yet he never denied that he was measuring emotions. Ekman did not hesitate to address internal states, and in fact he believed that facial expressions could not be understood without acknowledging emotions as their source. Emotions rarely stay on the inside, he said, because "one of the most distinctive features of an emotion is that it is typically not hidden: we hear and see signs of it in the expression."[2]

You'd think that since Ekman worked on our own species, he had nothing to worry about. But sadly, academics get into the weirdest battles, which we often neither understand afterward nor even remember. This happened with human facial expressions, which were considered trivial, unworthy of attention, or so immensely variable across the globe that they were best regarded as cultural artifacts. Tying them to biology, as Ekman was trying to do, was doomed from the start. All this changed, however, when Ekman visited the head of the resistance, an anthropologist who insisted that human emotions and their expressions are infinitely malleable. Expecting to find cabinets full of field notes, films, and photographs of human body language, Ekman asked if he could get a look at his records. To his astonishment, the answer was that none existed. The anthropologist claimed that all his data were in his head. This didn't look good: verifiable data are the bedrock of science. Could the entire cultural castle have been built on sand?

Ekman set up controlled tests with people from more than twenty different nations, showing them pictures of emotional faces. All these people labeled human expressions more or less the same way, showing little variation in recognizing anger, fear, happiness, and so on. A laugh means the same all over the world. One possible alternative explanation bothered Ekman, though. What if people everywhere were affected by popular Hollywood movies and television shows? Could this account for the uniformity of reactions? He traveled to one of the farthest corners of the planet to administer his tests to a preliterate tribe in Papua New Guinea. Not only had these people never heard of John Wayne or Marilyn Monroe, they were unfamiliar with television and magazines, period. Yet they still correctly identified most of the emotional faces that Ekman held in front of them, and they themselves showed no novel, unusual expressions in one hundred thousand feet of motion pictures of their daily lives. Ekman's data so powerfully argued in favor of universality that they permanently altered our view of human emotions and their expression. Nowadays, we consider them part of human nature.[3]

We should realize, though, how much all these studies rely on language. We are comparing not just faces and how we judge them but also the labels we attach to them. Since every language has its own emotional vocabulary, translation remains an issue. The only way around it is direct observation of how expressions are being used. If it is true that the environment shapes facial expressions, then children who are born blind and deaf should show no expressions at all, or only strange ones, because they've never seen the faces of people around them. Yet in studies of these children, they laugh, smile, and cry in the same way and under the same circumstances as any typical child. Since their situation excludes learning from models, how could anyone doubt that emotional expressions are part of biology?[4]

We have thus returned to Charles Darwin's position in his 1872 book *The Expression of the Emotions in Man and Animals*. Darwin stressed that facial expressions are part of our species's repertoire and pointed out similarities with monkeys and apes, suggesting that all primates have similar emotions. It was a landmark book—acknowledged by everyone in the field today—but it is the only major book by Darwin that, after its initial success, was promptly forgotten, then overlooked for almost a century before we returned to it. Why? Because hard-core scientists felt his language was too free and anthropomorphic. It embarrassed them when he wrote of a cat's "affectionate frame of mind" when she rubs against your leg, a chimpanzee's "disappointed and sulky" pouting of his lips, and cows that "frisk about from pleasure" while ridiculously throwing up their tails. What nonsense! Moreover, his suggestion that we convey our own noble sensibilities through facial movements that we share with "lower" animals was roundly insulting.

Among all the similarities, however, Darwin also noted exceptions. Both blushing and frowning, he thought, might be uniquely human. For blushing, he was absolutely right. I don't know of any quick face-reddening in other primates. Blushing remains an evolutionary mystery, especially for cynics who insist that the whole point of social life is the selfish exploitation of others. If this were true, wouldn't we be much better off without blood uncontrollably rushing to our cheeks and neck, where the change in skin color stands out like a light tower? If blushing keeps us honest, we'd need to wonder why evolution equipped us and no other species with such a conspicuous signal. Or as Mark Twain put it: "Man is the only animal that blushes—or needs to."

With regard to frowning, on the other hand, Darwin was only partly correct. He cited a contemporary expert who thought it was a peculiarly human reflection of superior intelligence, because the frown "knits the eyebrows with an energetic effort, which unaccountably, but irresistibly, conveys the idea of mind."[5] We have no reason, however, to pound our chests over a tiny muscle near the eyebrows. We now know it is present in other species. Wanting to explore its effect on nonhuman faces, Darwin made several visits to the Zoological Society's Gardens, in London. In a letter to his sister, he described his encounter with Jenny the orangutan:

> *I saw also the Ourang-outang in great perfection: the keeper showed her an apple, but would not give it her, whereupon she threw herself on her back, kicked & cried, precisely like a naughty child. She then looked very sulky & after two or three fits of pashion, the keeper said, "Jenny if you will stop bawling & be a good girl, I will give you the apple." She certainly understood every word of this, &, though like a child, she had great work to stop whining, she at last succeeded, & then got the apple, with which she jumped into an arm chair & began eating it, with the most contented countenance imaginable.*[6]

Because Darwin assumed that concentrating apes, like concentrating people, would frown when they were frustrated, he tried to aggravate Jenny and other apes into frowning by giving them a nearly impossible task. But they never frowned while tackling the problem. Ever since then, scientists have suggested that frowning might be uniquely human, but in fact apes can and do frown, as Darwin found out by tickling their noses with a straw: it made them crumple up their faces while "slight vertical furrows appeared between the eyebrows."[7] Chimps and orangutans have prominent bony eyebrow ridges that shield their eyes and make frowning difficult for them to do and hard for others to detect. But bonobos, with their flatter, more open face, frown easily, and they do so at the same moments as we do. While warning another one, for example, bonobos will narrow their eyes in a piercing stare with knitted brows, which looks exactly like the angry glower of our species.

I also distinctly remember such a stare by a chimpanzee. It was by one of my favorite old ladies, Borie, who had not only daughters but also grandchildren in the colony at the Yerkes Field Station. On an especially hot Georgia day, I had taken the hose to offer water to the apes. Fresh water is of course permanently available to them, but much as city kids love sprinklers, chimps find it much more fun to drink from a water hose that delivers large quantities. A dozen apes were pushing one another aside with their mouths wide open to catch the cool water. Then one of the little ones gave a sharp yelp when he got sprayed. No one cared much, but Borie immediately ran up to me and gave me an angry stare, warning me to be more careful. Up close, her intense frown was unmistakable.

The best way to understand animal emotions is just to watch spontaneous behavior, either in the wild or captivity. Students of animal behavior have documented hundreds, even thousands, of instances of the use of an expression. This is how we know that apes laugh during play, and that while munching on their favorite food, they give special grunts that invite everyone around them to join the feast. We record the events leading up to the expression as well as how it affects others. Does a given signal start a fight, stop a fight, or pave the way for reconciliation? We have entire catalogs, known as ethograms, of the typical signals of every species, not just primates but also horses, elephants, crows, lions, chickens, hyenas, and so on. One of the first ethograms concerned the wolf, including all its tail movements, ear positions, hair raising, vocalizations, baring of teeth, and so on. Ethograms can be quite elaborate, indicating a rich repertoire. For mice and rats, we have them, too.

Rodent faces were long thought to be unaffected by emotions, but detailed studies show that they express anguish through narrowed eyes, flattened ears, and swollen cheeks. Other rodents have no trouble recognizing these faces, because in experiments they prefer to sit close to a photograph of a rat with a relaxed face rather than one revealing pain. Conversely, rats also share good feelings. When Swiss scientists set up a positive treatment program consisting of daily tickling and play with lab rats, they analyzed their faces at a quiet moment after every session. They could tell which rats had received the positive treatment just from

| Joy | Anger | Shame | Love |
| Sad | Lust | Hunger | Fear |

Since rodents were long thought to have immobile faces, people made fun of them, as in this cartoon, which features identical faces said to express different feelings. But the joke is on us, now that we know that rats and mice signal both pain and pleasure in their faces.

looking at them, because of their pinker and more relaxed-looking ears. These studies put an end to the idea—mocked in cartoons of rats with identical poker faces labeled as different emotions—that the rodent face is static.[8]

One of the most expressive faces on the planet moves around on four hooves. That horses, donkeys, and zebras have a rich facial palette is perhaps unsurprising given how social and visual these animals are. The Equine FACS (Ekman's Facial Action Coding System applied to horses) recognizes no less than seventeen distinct muscle movements, produced in countless combinations. Horses snort more when they are content, greet each other by pulling back their lip corners, curl their upper lip during *Flehmen* (when they pick up an unusual scent), show white sclera while opening their eyes wide in fear, and have a great variety of ear positions.[9] Anyone who has a dog or a cat at home knows that ears are incredibly effective signaling devices, so much so that I consider humanity's immobile ears a serious handicap.

Dogs, too, have been studied for how they produce and perceive faces, including ours. We have concluded that they communicate intentionally from the fact that their faces change more in response to a watching human face than one who has turned his back. One common dog expression occurs when they pull their inner eyebrow, which enlarges their eyes. We fall for the cuteness of rounded faces with big eyes, a sensitivity massively exploited by animated movies. Dogs' inner-brow pull makes them look sadder and more puppy-like, which even affects pet adoptions. Observers at shelters have noticed that dogs who direct this face at human visitors are rehomed more easily than those who fail to do so. Clearly, man's best friend knows how to pull our emotional strings.[10]

People generally like to focus on expressions we share with other species, and the primates naturally are prime examples. This is where Jan comes in as the world's leading expert. In the 1970s, he conducted observations in far greater detail than anyone before him, making minute comparisons of how baboons rapidly smack their lips, or how male pigtail macaques raise their chin with pursed lips to court a female. Jan's main topic, however, was the laugh and how it differs from the smile.

Equine faces are about as expressive as those of the primates. Here a pony shows the *Flehmen* response, which is typical when encountering a new smell, or after a stallion has picked up the scent of a mare's urine. Rolling back the upper lip helps move the scent to receptors in the vomeronasal organ. Cats show a similar grimace when they come upon an unusual smell.

Even though the two expressions are often put on the same scale—as if the smile is a low-intensity laugh—Jan demonstrated that they hail from separate origins.[11]

From Ear to Ear

I can't stand TV sitcoms and Hollywood movies featuring monkeys and apes: every time I see a dressed-up simian actor produce one of their silly grins, I cringe. People may think they're hilarious, but I know their mood is the opposite of happy. It's hard to get these animals to bare their teeth without scaring them—only punishment and domination can call forth these expressions. Behind the scenes, a trainer is waving his electric cattle prod or leather whip to make clear what will happen if the animals fail to obey. They are terrified! This is why apes in the movies are almost never adult: grown apes are far too strong for human trainers to dominate, and they're much wilier than any of the large cats. Only younger apes can be intimidated to the point that they will grin on command.

Many questions surround the bared-teeth grin, such as how this toothy expression became a friendly one in our species and where it came from. The latter question may seem odd, but everything in nature is a modification of something older. Our hands came from the fore-limbs of land vertebrates, which derived from the pectoral fins of fish. Our lungs evolved out of fish bladders.

With regard to signals, too, we wonder about their origin. The process of transformation from earlier versions is known as *ritualization*. For example, we mimic the instrumental act of holding an old-fashioned phone by extending the thumb and pinky finger and holding them against our ear: this gesture has been transformed into a "call me!" signal. Ritualization does the same, but on an evolutionary scale. The woodpecker's irregular tapping on a tree in order to find grubs became rhythmic drumming on hollow logs to announce a territory. And the soft chewing sounds that monkeys make when picking lice and ticks out of each other's hair became a friendly greeting with lifted eyebrows and audible lip-smacking, as if to say "I'd love to groom you!"

The bared-teeth grin is not to be confused with the flashing of teeth with a wide-open mouth and intense staring eyes. That fierce face, which looks like an intention to bite, acts as a threat. In a grin, in contrast, the mouth is closed and the lips are retracted to expose the teeth and gums. The row of bright white teeth makes it a conspicuous signal, visible from far away, yet its meaning is the exact opposite of a threat. The expression derives from a defensive reflex.[12] We automatically pull our lips back from our teeth when we peel a citrus fruit, for example, which risks spraying acid drops into our face.

Fear and unease also pull at our mouth corners. Films of people riding in roller coasters show an enormous amount of grinning, not delighted smiles but terrified grimaces. The same happens in other primates. I once went baboon watching on the plains of Kenya during a drought. The baboons consumed tons of acacia beans, which made my downwind journey following a hundred flatulent monkeys rather smelly. They often stopped to feast on succulent cactus, an invasive plant that they normally leave alone because of its sharp spines. Before sinking their teeth into the cactus, the monkeys would draw their lips far back to avoid perforating them. Their reason was practical, but their faces featured the same grin as in social exchanges, where it acts as a submissive signal.

In a rhesus monkey group that I studied, the mighty alpha female, Orange, needed only to walk around, and all the females she passed would flash grins at her—especially if she walked in their direction, and even more so if she honored their huddle by joining them. A dozen grinning faces might be staring at Orange. None of the other females would move out of her way, because the whole point of the expression is to stay put while at the same time showing respect. The other females essentially told Orange, "I'm subordinate, I'd never dare challenge you."

Orange was so secure in her position that she rarely needed to use force, and by showing their teeth, the other females removed any reason she might have had for throwing her weight around. Among rhesus monkeys, this expression is 100 percent unidirectional: it is given by the subordinate to the dominant, never the other way around. As such, it is an unambiguous marker of the hierarchy. Every species has signals for this purpose. Humans signal subordination by bowing, groveling, laughing at the boss's jokes, kissing the don's ring, saluting, and so on. Chimpanzees lower themselves in the presence of high-ranking individuals and issue a special kind of grunt to greet them. But the original primate signal to make clear that you rank below someone else is a grin with the mouth corners pulled back.

But far more underlies this expression than fear. When a monkey is simply scared, such as when it spots a snake or predator, it freezes (to avoid detection) or else it runs away as fast as possible. This is what plain fear looks like. No grin is involved, as it doesn't fit these situations. The grin is an intensely social signal that mixes fear with a desire for acceptance. It is a bit like the way a dog may greet you, with flattened ears and tucked-in tail, while rolling on his back and whining. He exposes his belly and throat while trusting that you will not use weaponry on his most vulnerable body parts. No one would mistake the canine rollover for an act of fear, because dogs often behave this way while approaching the other as an opening move. It can be positively friendly. The same applies to the monkey grin: it expresses a desire for good relations. Hence Orange received the signal many times a day, whereas a snake never would.

I once befriended a young rhesus monkey, Curry, who lived in the same troop as Orange in a large outdoor area. It was surrounded by a mesh fence, through which I'd take pictures. As I stood there day in and day out, the monkeys got used to me. Initially, of course, they threatened me and tried to grab my camera, but they ended up ignoring me, which made my life as a photographer much easier. Curry sought me out by the mesh, often showing her teeth in an act of submission while approaching me. She liked to sit close by, sometimes holding on to one of my fingers through the mesh with her little hand. I had to be careful, because monkeys bite, but Curry could be trusted. Since she was low-ranking, she may have gained safety from hanging out with me. Every time I looked in her direction, she flashed her teeth, but that's because eye contact is threatening to a macaque. Her name was apt, as she grinned to curry favor with me.

The great apes go a step further: their grin, although still a nervous signal, is more positive. In many ways, their expressions and the way they use them are more like ours. Bonobos sometimes bare their teeth in friendly and pleasurable situations, such as in the midst of sexual intercourse. One German investigator spoke of an *Orgasmusgesicht* (orgasm face) given by females while they stare into their partner's face—bonobos often mate face to face. The same expression may be used to calm down or win over others and not purely along hierarchical lines, as in the monkeys. Dominant individuals also bare their teeth when they try to reassure others. For example, when an infant wanted to steal a female's food, she dealt with it by gently moving the food out of his reach while flashing a big grin from ear to ear. This way she prevented a tantrum. Friendly grins are also a way to smooth things over when play gets too rough. Only rarely do apes lift up their mouth corners during a grin, but if they do, it looks exactly like a human smile.

Inasmuch as an ape's grin betrays anxiety, it isn't always welcome. Male chimpanzees—who are always in the business of trying to intimidate one another—don't like to reveal anxieties in the presence of a rival. It's a sign of weakness. When one male hoots and puts up his hair while picking up a big rock, it may cause unease in another because it announces a confrontation. A nervous grin may appear on the target's face. Under these circumstances, I have seen the grinning male abruptly turn away so that the first male can't see his expression. I have also seen males hide their grin behind a hand, or even actively wipe it off their face. One male used his fingers to push his own lips back into place, over his teeth, before turning around to confront his challenger. To me, this suggests that chimpanzees are aware of how their signals come across. It also shows they have better control over their hands than over their faces. The same is true for us. We may be capable of producing expressions on command, but it's hard to change one that comes up involuntarily. To look happy when you are angry, for example, or to look angry when in reality you are amused (as may happen to parents with their children), is nearly impossible.

The human smile derives from the nervous grin found in other primates. We employ it when there is a potential for conflict, something we are always worried about even under the friendliest circumstances. We bring flowers or a bottle of wine when we are invading other people's home territory, and we greet each other by waving an open hand, a gesture thought to originate from showing that we carry no weapons. But the smile remains our main tool to improve the mood. Copying another's smile makes everyone happier, or as Louis Armstrong sang: "When you're smiling, the whole world smiles with you."

Reprimanded children sometimes can't stop smiling, which risks being mistaken for disrespect. All they're doing, though, is nervously signaling nonhostility. This is why women smile more than men, and why men who smile are often in need of friendly relations. One study explicitly looked at this underdog quality of the smile in pictures taken right before matches in the Ultimate Fighting Championship. The photographs show both fighters defiantly staring at each other. Analysis of a large number of pictures revealed that the fighter with the more intense smile was the one who'd end up losing the fight later that day. The investigators concluded that smiling indicates a lack of physical dominance, and that the fighter who smiles the most is the one most in need of appeasement.[13]

I seriously doubt that the smile is our species's "happy" face, as is often stated in books about human emotions. Its background is much

richer, with meanings other than cheeriness. Depending on the circumstances, the smile can convey nervousness, a need to please, reassurance
to anxious others, a welcoming attitude, submission, amusement, attraction, and so on. Are all these feelings captured by calling them "happy"?
Our labels grossly simplify emotional displays, like the way we give each
emoticon a single meaning. Many of us now use smiley or frowny faces
to punctuate text messages, which suggests that language by itself is not
as effective as advertised. We feel the need to add nonverbal cues to prevent a peace offer from being mistaken for an act of revenge, or a joke
from being taken as an insult. Emoticons and words are poor substitutes
for the body itself, though: through gaze direction, expressions, tone of
voice, posture, pupil dilation, and gestures, the body is much better than
language at communicating a wide range of meanings.

Nevertheless, we keep simplifying the body's messaging system by
pairing still pictorial representations of it with sadness, happiness, fear,
anger, surprise, or disgust, known as the "basic" emotions. Never mind
that most emotional states are a blend of separate tendencies. When I was
a boy, I went up to the roof of our house to train to become a helper to
the Dutch St. Nicholas, a bearded bishop who drops presents through
chimneys. Obviously he can't do this job on his own. Not realizing that
getting onto a roof is much easier than getting off, I got stuck. My dad
scolded me when I was discovered in my precarious situation. His reaction looked very much like anger, with threatening gestures, raised voice,
and a purple face. But his anger was triggered by apprehension and was
mixed with hope that some good discipline might keep me from being
so stupid again. It sure did! My point here is that every display of emotion needs to be judged in a wider context. A single label rarely suffices.
To call my father's state "angry" fails to do it justice without also mentioning love and worry.

The same desire for simplification applies to animal emotions, perhaps even more so, because we like to think that their emotions must
be simpler than ours. In fact, *The Oxford Companion to Animal Behaviour,*

published in 1987, maintained that there was absolutely no point studying animal emotions, because they tell us nothing new, and besides, "animals are restricted to just a few basic emotions."[14] Absent a science of animal emotions, one wonders how the author ever reached this conclusion. It is a bit like the old claim, repeated over and over in the literature, that we have hundreds of muscles in our faces, far more than any other species. Following the *scala naturae* view, it was assumed that the closer an animal gets to us on the evolutionary ladder, the richer its palette of emotions must be, and therefore the more varied its facial musculature.

But there is really no good reason why this should be so. When a team of behavioral scientists and anthropologists finally tested the idea by carefully dissecting the faces of two dead chimpanzees, they found the exact same number of mimetic muscles as in the human face—and surprisingly few differences.[15] We could have predicted this, of course, because Nikolaas Tulp, the Dutch anatomist immortalized in Rembrandt's *The Anatomy Lesson*, had long ago reached a similar conclusion. In 1641 Tulp was the first to dissect an ape cadaver and found that it resembled the human body so closely in its structural details, musculature, organs, and so on, that the species looked like two drops of water.

Despite these similarities, the human smile differs from the ape equivalent in that we typically pull up our mouth corners and infuse the expression with even more friendliness and affection. This applies only to the real smile, though. We often wear plastic smiles with no deep meaning whatsoever. The smiles of airplane personnel and smiles produced for cameras ("say cheese!") are artificial, for public consumption. Only the so-called Duchenne smile is a sincere expression of joy and positive feeling. In the nineteenth century, the French neurologist Duchenne de Boulogne tested facial displays by electrically stimulating the face of a man who lacked pain perception. Duchenne produced and photographed all sorts of expressions this way, but the man's smiles never looked happy. In fact, they looked fake. One time Duchenne told

Our species has two kinds of smile. The full version is known as the Duchenne smile, named after the French neurologist who pioneered the study of facial expressions. He discovered that lip retraction and upturned mouth corners are not enough. In the smile on the left, the muscles around the eyes are pulled together, causing wrinkles while narrowing the eyes: the Duchenne smile. In the face on the right, the eyes fail to join the smiling mouth, making for a fake smile.

the same man a funny joke and triggered a much better smile, because instead of just smiling with his mouth, as he had been doing thus far, he now narrowed the muscles around his eyes as well. Duchenne perceptively concluded that while the mouth can produce a smile on command, the muscles near the eyes don't obey as well. Their contraction completes a smile to indicate genuine enjoyment.

And there you have it. Some smiles are mere signals to the rest of the world, produced deliberately and found all over the Internet in portraits of politicians and celebrities and in millions of selfies. Others arise from a specific inner state, as sincere reflections of enjoyment, happiness, or affection. These smiles are much harder to feign.

That our faces most of the time mirror true feelings may seem obvious enough, but even this simple idea was once controversial. Scientists strenuously objected to Darwin's use of the term *expression* as too suggestive, as implying that the face conveys what's going on inside. Even though psychology literally is the study of the *psyche*—Greek for "soul" or "spirit"—many psychologists didn't like references to hidden processes and declared the soul off limits. They preferred to stick to observable behavior, and regarded facial displays as little flags of different colors that we wave in order to alert those around us to our future behavior.

Darwin won this battle too, because if our facial expressions were mere flags, we should have no trouble choosing which ones to wave and which to leave folded. Every facial configuration would be as easy to summon as a fake smile. But in fact, we have far less control over our faces than over the rest of our bodies. Like the chimpanzees, we sometimes hide a smile behind a hand (or a book, or a newspaper) because we're simply unable to suppress it. And we regularly smile, or shed tears, or pull a disgusted face while we are unseen by others, such as when we are talking on the phone or reading a novel. From a communication perspective, this doesn't make any sense. We should have completely blank faces while talking on the phone.

Unless, of course, we evolved to communicate inner states involuntarily. In that case, expression and communication are the same thing. We don't fully control our faces because we don't fully control our emotions. That this allows others to read our feelings is a bonus. Indeed, the tight link between what goes on inside and what we reveal on the outside may well be the whole reason why facial expressions evolved.

That Was Funny!

I once attended a lecture by a philosopher who was baffled by the nonverbal aspects of human communication. He preferred the written and spoken word, but could of course not get around all the faces we pull and the gestures we make. Why do we need all these accompaniments, he wondered, and especially, why are they so exaggerated? When we laugh at a joke, for example, we lose partial control over our bodies and produce an enormous amount of *ha ha ha!* noise that can be heard far and wide. Why can't we just calmly say "That was funny!" and leave it at that?

I imagined a stand-up comic in a small theater telling the greatest joke of all time, but instead of people falling out of their chairs roaring with laughter, they'd all remain quietly in place muttering "That was funny!" The comic, knowing that humanity's noble sense of humor is irrevocably married to something much more animalistic, would obviously feel deeply offended. Laughter shows how central the body is to our existence, including our mental life. Laughter brings body and mind together, fusing them into a single whole. We may experience this as a loss of control because we like the mind to be the boss. As the theater critic John Lahr put it, "To watch inspired laughter register with an audience is to be present at a great and violent mystery. Faces convulse, tears stream, bodies collapse, not in agony but in rapture."[16]

When we laugh, we go crazy. We become limp, we lean on each other, we turn red, and we shed tears to the point of dissolving the dividing line with crying. We literally pee in our pants! After an evening of laughter, we are totally exhausted. This is partly because intense laughter is marked by more exhalations (producing sound) than inhalations (taking in oxygen), so we end up gasping for air. Laughter is one of the great joys of being human, with well-known health benefits, such as stress reduction, stimulation of heart and lungs, and release of endorphins. Nevertheless we should hope that extraterrestrials never get to watch a group of out-of-control laughing humans, because they'd probably abandon the idea of having found intelligent life.

Humor is not always the trigger for laughter. When psychologists unobtrusively take notes on human behavior in shopping malls and on the sidewalks of our natural habitat, they find that the majority of laughs occur after mundane statements that are anything but amusing. Try it yourself. Notice when people laugh in spontaneous chit-chat, and you'll see that often it's about nothing at all—no joke, no pun, no odd remark. It's just a laugh inserted in the flow of conversation, usually echoed by the partner. Humor is not central to laughter: social relationships are. Our supernoisy barklike displays announce mutual liking and well-being. The laughter of a group of people broadcasts solidarity and togetherness, not unlike the howling of a pack of wolves.[17]

The loud volume of our species's laughter gets me every time: apes laugh much more softly, and monkeys can hardly be heard at all. My guess is that the loudness is inversely proportional to predation risk. If the laughter of the young of other primates were as earsplitting as the laughter of our children in schoolyards, predators would have no trouble locating them and pouncing at the right moment. Humans can afford to be noisy, although we obviously also do a lot of soft chuckling and snickering.

At his eightieth birthday party, Jan gave a splendid demonstration of the human laugh sequence: he emitted a series of loud belly *ha*'s, then a deep inhalation, drawn out for extra effect. The room burst into laughter, not only because this sequence is a signature of our species, but also because it is incredibly contagious. In experiments, humans automatically mimic laughing faces displayed on a computer screen, and the whole purpose of adding laugh tracks to sitcoms is to induce contagion. Detailed video analyses of ape behavior finds similar mimicry. When one juvenile orangutan approaches another with a laughing face, the other instantly adopts the same expression, which is why typically both play partners laugh rather than just one of them.[18] Even birds display this contagious behavior. New Zealand parrots, known as keas, instantly turn playful when their species's warbling play vocalizations resound from a hidden speaker. The calls, which resemble laughter a little, affect their mood. The keas immediately invite other birds to play, pick up toys to manipulate, or engage in aerial acrobatics. Nothing is as infectious as playfulness and laughter.[19]

The repetitiveness of the primate laugh derives from rhythmic panting. In apes, laughing starts with audible panting, which grows more and more vocal the more intense the encounter becomes. By itself, separate from play, rapid panting expresses relief, joy, and a desire for contact, as when a female chimpanzee walks up to her best friend and utters audible pants before kissing her. Just so, Mama would rapidly pant at me before grasping my arm, then splutter and smack when she groomed me. Working with apes, you learn to be careful and watch their signals. All these soft sounds indicated good intentions, so much so that without them I might have been reluctant to let Mama grab my arm.

Nadia Ladygina-Kohts, the Russian scientist who a century ago compared the emotional development of her juvenile chimpanzee, Joni, with that of her own little son, offered examples of joyful moments that triggered panting. One day Joni saw Kohts leave the house and started to whine, but as soon as she changed her mind and stayed, he hurried over to her with rapid pants. When Joni expected a serious scolding for a misdeed, but was in fact treated cordially, he panted in appreciation. This rapid panting, which communicates joy and positive feelings, became the basis of laughter, which communicates the same but much more loudly.[20]

Animal play can be rough, as players may wrestle, gnaw, jump on top of each other, and drag each other around. Without an unambiguous signal to clarify their intentions, play behavior might be mistaken for a fight. Play signals tell others that they have nothing to worry about, that none of this is serious. For example, dogs may "play bow" (crouch down on their forelimbs with their butt in the air) to help set play apart from conflict. But as soon as one dog misbehaves and accidentally bites the other, play ceases abruptly. A new play bow will be required as "apology," so the victim can overlook the offense and resume play.

Laughter serves the same purpose: it puts other behavior into context. One chimp pushes another firmly to the ground and puts his teeth in her neck, leaving her no escape, but since both utter a constant stream of hoarse laughs, they stay totally relaxed. They know that this is just for fun. Since play signals help interpret other behavior, they are known as *meta*-communication: they communicate about communication.[21] Similarly, if I approach a colleague and slap him on the shoulder with a laugh, he will perceive it quite differently than he would if I did so without a sound or without any expression on my face. My laugh delivers a meta-signal about the hand that hit him. Laughing reframes what we say or do and takes the sting out of potentially offensive remarks, which is why we use it all the time, even when nothing particularly amusing is going on.

Laughter signals not only playfulness between buddies but also to the outside world. When others see or hear laughing, they know everything is fine. Chimpanzees are smart enough to employ laughs this way. We once analyzed hundreds of wrestling matches among young chimps to see at which moments they laughed. We were particularly interested in youngsters far apart in age, as their games often got too rough for the younger one. As soon as this happened, the mother of the youngest partner would step in, sometimes hitting his playmate over the head. It was always the fault of the older one! We found that when juveniles play with infants, they laugh far more when the infant's mother is watching them than when they are alone. Under the eyes of a protective mother, laughter projects a jolly mood as if to say, "Look how much fun we're having!"[22]

If a clique of people laugh, and you are not part of it, you will feel excluded. Laughter often emphasizes the in-group at the expense of the out-group. It is such a powerful form of taunting and teasing that some have proposed that hostility is at its root. These theories speak of "ostracizing humor" aimed at outsiders or people of a different race, and depict laughter as a malicious act.[23] The sixteenth-century English philosopher Thomas Hobbes, for example, saw laughter as an expression of superiority, as if the whole purpose of human joking were to make fun of others. What a miserable life this man must have led!

Laughter is far more typical of affectionate relations between buddies, lovers, spouses, parents and children, and so on. Where would marriages be without the essential glue of humor? I'm from a big family and remember with fondness the laughter around the dinner table, which could get so intense that I felt like dying. I'd have to exit the room to catch my breath and regain composure. The earliest laughter in our lives always occurs in a nurturing context, as it does in the other primates. A gorilla mother tickles the belly of her tiny baby with her big finger just a few days after birth, producing the very first laugh. In our own species, mothers and babies have lots of exchanges, in which they pay attention to every shift in each other's expression and voice, with ample smiling and laughter. This is the original context, which is totally lacking in malice.

Physical stimulation remains part of it, and it must have a long evolutionary history, because tickling is also connected with laughlike sounds in rats. The late Estonian-American neuroscientist Jaak Panksepp did more than anyone else to pioneer animal emotions as an acceptable topic of discussion. Panksepp was initially ridiculed for the whole idea of laughing rats. These rodents remain despised and underestimated, but since I once had them as pets, I have no doubts whatsoever that they are complex animals that bond and play. Panksepp noticed that rats like being tickled by human fingers, so much so that they will come back for more. When you withdraw your hand and move it somewhere else, they will follow, seeking out stimulation while uttering bursts of 50-kHz chirps that are above the human hearing range.

One anonymous rat fancier tried it at home:

I decided to do a little experimenting of my own with my son's pet rat, Pinky, a young male. Within one week, Pinky was completely conditioned to playing with me and every once in a while even emits a high pitched squeak that I can hear. The second I walk into the room, he starts gnawing on the bars of his cage and bouncing around like a kangaroo until I tickle him. He tackles my hand, nibbles, licks, rolls over onto his back to expose his tummy to be tickled (that's his favorite), and does bunny kicks when I wrestle with him.[24]

Panksepp concluded that for rats, tickling is a rewarding experience (hence the chasing of the hand) that requires the right mood. If the animals are anxious or frightened by cat smell or bright lights, no amount

Much as humans and apes laugh while being tickled, rats produce high-pitched chirps—outside the range of human hearing. They actively seek out a human hand that tickles them, indicating they find it pleasurable.

of tickling will induce laughter. Their enthusiasm also depends on their previous experience and familiarity, because rats more eagerly approach a hand that has tickled them, while uttering high-pitched chirps, than a hand that has only petted them. The rats make little frolicsome movements, known as "joy jumps," that are typical of all playing mammals, including goats, dogs, cats, horses, primates, and so on. Darwin's frisky cows come to mind. Even though animals have all sorts of play signals, the one constant is abrupt random jumpiness. They dance toward you with an arched back (cats) or turn around their axis and leap onto the forbidden couch (dogs) to show how ready they are for a chase. The joy jump is so recognizable that it is easily understood between species. In captivity, a rhino calf may play with a dog, or a dog with an otter, or a foal with a goat, and in the wild young chimpanzees have been observed wrestling with baboons, and ravens and wolves teasing each other. Play has its own universal language.

We may use laughter to defuse an awkward or tense situation. That is less common in other species but not excluded. Among chimpanzees, I have seen males take the heat out of potential conflict. Three adult males, with all their hair up, had just been performing impressive charging displays. This is a very tense, potentially dangerous situation in which rivals test each other's nerves. They swing from branch to branch, dislodge heavy rocks, throw things around, and bang on resounding surfaces. But when these three males walked away from the scene, all of a sudden one of them literally pulled the leg of another. This male resisted and tried to free his foot, all the while laughing. Then the third male joined the act, and before long three big males were galloping around, punching each other in the sides and uttering hoarse laughs, while literally letting their hair down. The tension had been broken.

Aristotle thought laughter was what set humans apart from beasts, and many psychologists still doubt that any animal laughs for joy or because something is funny. It is well known, though, that apes love slapstick movies, probably because of all the physical mishaps. When a person they like walks toward them and slips or falls, their first reaction is worried tension, but if the person turns out to be fine, they laugh with apparent relief, the way we do under similar circumstances. I have already described Mama's laughing when she discovered that she had been tricked by a human with a panther mask. Similar reactions can be seen in bonobos. Long ago the bonobo enclosure at the San Diego Zoo had a deep, dry moat to separate the apes from the public. On the bonobo side, a plastic chain hung down into the moat so that the apes could descend and come back up whenever they liked. When the alpha male, Vernon, went down, however, an adolescent male, Kalind, sometimes quickly pulled up the chain after him. Vernon would be stuck, while Kalind would look down at him with a wide-open laughing face while slapping the side of the moat. He was making fun of the boss. The only other adult present would usually rush to the scene to rescue her mate by dropping the chain back down, and she would stand guard until he climbed out.

Another amused laugh was videotaped by Japanese fieldworkers in West Africa. A nine-year-old wild chimpanzee was happily smashing nuts with stones using a common hammer-and-anvil technique. One by one, he'd put the tough palm nuts on the flat surface of a large stone while holding a smaller one in his other hand, then hit them with it until they cracked. In the forest, it is not easy to find the right combination of stones for this task. The male's mother glanced at his perfect tools before walking up and grooming him. This is often an invitation for return grooming, so when she finished, she stood there waiting for him to spin around and groom her. Doing so, he left his stones unattended, and within a few seconds his mom reached over and snatched them. It looked intentional, as if her approach and brief grooming had been a way to distract him. The very moment she collected his tools, you could hear and see her softly laugh to herself, delighted that her little scheme had worked.[25]

Admittedly, this evidence is anecdotal, but these incidents hint that ape laughter may be more than just a play signal. Occasionally, it seems to approach the broader meaning of glee, bonding, and tension breaking that we know of our own species.

Blended Emotions

The evolutionary histories of the laugh and the smile show how right Jan was to propose separate origins for them. They originate from different corners of the emotion spectrum. One started as an expression of fear and submission, which became a sign of nonhostility and ultimately affection. The other started as a play indicator sensitive to roughhousing and tickling, then turned into a signal of bonding and well-being, even fun and happiness. Both expressions have been growing closer in our species, and since we often blend emotions, they ended up merging. We often move from a smile to a laugh and vice versa, or we show mixtures of the two.

Blended expressions are typical of the hominids, the small primate family of humans and apes. Whereas most other animals, including monkeys, have discrete calls and displays, the hominids stand out for their graded communication. A monkey will give a threat face, a bared-teeth grin, or a play face, but not a combination or mixture of these expressions. Its signals are fixed and stereotypical, quite separate from one another, as if they were either blue or red, never purple. This is a serious limitation compared to the apes, which easily move back and forth between a pout face, a whimper, and a bared-teeth yelp. Their faces are in constant motion to cover a wide range of tendencies, even if they are in conflict. In the same way, a child may cry, then laugh through her tears, then cry some more.

Using a classification of twenty-five facial expressions, we analyzed literally thousands of them in the Yerkes chimpanzee colony during daily life in their outdoor enclosure. We noticed an enormous degree of grading and blending.[26] For example, a young male seeking contact with the alpha male is afraid and sits at a distance waiting for a convivial sign. The young male gives friendly signals, such as holding out a hand to his leader with rapid pants, but also submissive grunts to show his respect. Or a female is interested in another's juicy watermelon but is upset when she keeps getting rebuffed and hesitates between continued begging and loud protest, which might trigger a fight. She moves between pouts and whines to request food and between yelps and soft screams that betray her mounting frustration. Social interactions are full of such conflicting tendencies, and human and ape faces reveal them all. They show not just a snapshot of one emotion or another but all the subtle shades in between. In fact, stand-alone emotional states are rare, which is why the whole goal of putting facial expressions into little boxes labeled "angry," "sad," or other basic emotions is so problematic. It works neither for us nor for our fellow hominids.

3 | BODY TO BODY

Empathy and Sympathy

M y very first experience with chimpanzees was in college at the Radboud University Nijmegen. To earn a few quick guilders, I took a position as research assistant in a psychology lab. On the first day, I heard that the job involved chimpanzees. This took me by surprise, because who in his right mind would keep apes on the top floor of a university building amid offices and classrooms? The living conditions were far from ideal and would never be permitted today, but I had a great time getting to know my two hairy friends.

Every day I tested them on the sort of cognitive tasks that might be perfect for rats but were not suitable for apes. In those days, psychologists still believed in universal laws of learning and intelligence; they had no interest in the special talents of each species. Not even brain size mattered to them. As the founder of the behaviorist school, B. F. Skinner, bluntly put it, "Pigeon, rat, monkey, which is which? It doesn't matter."[1] We now know, however, that there are many different kinds of intelligence, each one adapted to the special senses and natural history of a species. You simply cannot evaluate an ape or an elephant the way you evaluate a crow or an octopus. Apes, especially, are thinking beings who try to understand every problem they face. They lose interest once they have figured it out. Compared to a couple of rhesus monkeys tested in the same lab, our chimps underachieved, which goes to show that performance and intelligence are not the same thing. While the monkeys kept their eyes firmly on the rewards and settled on a routine to collect as many as they could, the apes got bored. The task was below their level. As a result, I spent much time roughhousing with them, which they liked much better.

This was how I first learned this species's typical sounds and other forms of communication, and also how to act like an ape, which is really not that hard given that humans are essentially apes. The only part I was unable to duplicate was their muscle strength. I couldn't swing hanging by a single finger from the ceiling or bounce between walls without touching the floor. Though not even six years old, they quickly grasped that I was a feeble being who didn't like being tied into the same knots that they applied to each other. I could slap them on their backs as hard as I could—so hard that any human would have burst out in angry protest—but they simply kept laughing as if this were the most hilarious thing I'd ever done.

Typical of their age, their sexual urges were emerging, which they were forced to project onto our species. Both males would have an erection as soon as they saw a woman walk by. They were so accurate in spotting the opposite sex that I wondered how they did it. Smell was unlikely, because their senses are like ours—vision is dominant. A male fellow student and I decided to test things out, leading to my very first behavioral experiment. We dressed up in skirts and wigs and raised our voices to see what kind of reaction we'd get. We walked in chatting and pointing at the chimps as if we were unexpected female visitors. They barely looked up. No erect penises, no confusion, except that they pulled at our skirts. A few minutes later one of the secretaries peeked around the corner, having seen two strange ladies walk by and thinking they were lost. With her, the chimps immediately showed the reaction we had been hoping for. We concluded that it is easier to fool people than chimps.

This experiment was almost like a prank. I'd hesitate to mention it except that it illustrates keen perception, which is the topic of this chapter. How does one organism read the body language of another? Many animals have the same sensitivity as those two chimps when it comes to distinguishing human genders. Even species quite distant from us, such as birds and cats, do so with ease. I have known many a parrot who liked only women or only men. They zoom in on the one visible sex difference that is found throughout the animal kingdom: male movements tend to be more brusque and resolute than those of females, which are more flowing and supple. We don't even need to see whole bodies to make this distinction. When scientists attached little lights to the arms, legs, and pelvis of people and filmed them walking, they found that these dots alone contain all the information we need to distinguish gender.[2] From watching just a few moving white specks against a dark background, subjects can tell right away if they are looking at a man or a woman. The walking pattern even varies with the stage of a woman's ovulatory cycle. If we can accurately judge people based on such scanty information, it is not hard to see why for many animals human maleness or femaleness is an open book. It also works in reverse, because from a distance, I can surely tell a male and female chimp apart based on the way they move.

Many years later we conducted a more scientific experiment on gender distinctions. It grew out of touchscreen research on face recognition, but ended with the discovery that chimpanzees are intimate with each other's derrieres. Sitting in front of a monitor, a chimp would first see a picture of a behind of its own species followed by two portraits. Only one portrait matched the behind they'd just seen—it showed the face of the same ape. The task was too easy if the faces were of different sexes, because male and female behinds are strikingly different, and the faces of both sexes differ, too.

But what if they had to choose between two male portraits after having seen a male behind, or between two female portraits after a female behind? Would they still pick the correct one? We found indeed that our chimps selected the portrait that went with the butt, but only with chimps that they knew personally. That they failed with strangers suggests that their choices were not based on something in the pictures, such as color or size, but on knowledge that came from outside, from seeing each other every day. Having a whole-body image of familiar individuals, they knew them so well that they could connect any part of their physique with any other part, such as the anterior with the posterior. We published our findings under the title "Faces and Behinds," and because everyone thought it was funny that apes could do this, we received an Ig Nobel award—a parody of the Nobel Prize that honors research that "first makes people laugh, and then think."[3]

Even though the same experiment has never been tried on humans—least of all unclothed ones—we must possess the same whole-body image, because we are all able to pick friends and relatives out of a crowd even if we only see their backs.

Wisdom of Ages

We perceive and interpret emotions using communication, empathy, and coordination and especially by reading body language. Because it's nearly impossible for researchers to study how people perceive emotions by observation alone, they gain the most knowledge from experiments, typically ones that present images on a touchscreen. Humans are tested like this all the time, but so are other species.

Our chimps get very excited by these studies, maybe because of their fascination with the immediate feedback a touchscreen provides, the way children are drawn to smartphones. In fact, the quickest way to get chimpanzees to enter our Cognition Building at Yerkes, which they do on a voluntary basis, is to move past their outdoor enclosure pushing a service cart carrying a computer. The chimps burst out hooting and run to the doors of the building where we do the testing, lining up to get in. They're eager to spend an hour of what they see as fun and games and what we see as cognitive testing. We don't even need to reward them for their performance: for them, touching images and solving puzzles are enjoyable in their own right. Some chimps become competitive about it: they hear from the monitor's sound how well they are doing (a correct solution produces a happier sound than an error), and they get upset if they hear a nearby companion doing better than they are. It's the best way to get them to focus!

I like experiments to be mutually enjoyable for both the scientists and the animals. The trick is to design interesting tasks. For example, for the longest time, we tested face recognition by showing primates human faces, then when they performed poorly, we concluded that only we, humans, recognize faces. Some scientists went so far as to claim the existence of a special face-recognition module in the human brain that evolved exclusively in our lineage. But then chimpanzees were tested on faces of their *own* kind. All of a sudden, they paid better attention and turned out to be just as good as humans.

They even showed signs of holistic perception. We humans don't recognize faces by how big the nose is or by how far apart the eyes are; rather, we take in the overall configuration, perceiving a face as a whole. The same turns out to be true of other primates, provided they are tested on their own kind. Even dogs—domesticated animals specifically bred to get along with us—are better at recognizing dog emotions than human ones. None of this is terribly surprising, but for the longest time we tested animals incorrectly based on the assumption that our faces must be the most distinct in the world. Clearly, neither apes nor dogs are as into us as we'd like them to be.

What about emotional expressions? Here it gets tricky, because we can't ask animals what their expressions mean. We can't give them a list of adjectives like *happy, sad,* and so on, the way Ekman did. Lisa Parr, then a student of mine, found an ingenious solution using physiological data. Physiology tells us how the body reacts, which is crucial, because emotions belong as much to the body as to the mind. The modern English word *emotion* derives from the French verb *émouvoir,* which translates as to "move," "touch," or "stir up." From further back in history, the Latin *emovere* means "to agitate." In other words, emotions can't leave us alone. They are mental states that make our hearts beat faster, our skin gain color, our faces tremble, our chests tighten, our voices rise, our tears flow, our stomachs turn, and so on.

Not only do emotions affect the body, the reverse is equally true. Emotions are heavily influenced by hormones (such as those of the menstrual cycle), sexual arousal, insomnia, hunger, exhaustion, sickness, and other bodily states. We associate different emotions with specific locations in the body, and the body in turn affects what we feel. For example, the enteric nervous system—a network of millions of neurons embedded in the lining of the digestive tract—may give us butterflies of anxiety in the pit of our stomach, which in turn tells our brain what we feel. Because of the enteric system's autonomy, it is also called our "second brain."

That emotions are rooted in the body explains why Western science has taken so long to appreciate them. In the West, we love the mind, while giving short shrift to the body. The mind is noble, while the body drags us down. We say the mind is strong while the flesh is weak, and we associate emotions with illogical and absurd decisions. "Don't get too emotional!" we warn. Until recently, emotions were mostly ignored as almost beneath human dignity.

Emotions often know better than we do what is good for us, even though not everyone is prepared to listen. While Charles Darwin was trying to decide whether to ask his cousin Emma Wedgwood to marry him, he drew up a long list of arguments in favor ("Object to be beloved

& played with—better than a dog anyhow") and against ("Not forced to visit relatives & to bend in every trifle").[4] This way he hoped to arrive at a perfectly rational decision, but I doubt very much that his list swayed him one way or the other. He even forgot the two points in favor of marriage that many of us would put at the top of the list: love and physical attraction. By writing down a firm QED (*quod erat demonstrandum*) that favored proposing to Emma, Darwin acted as if he had produced some sort of mathematical proof, but obviously his math was illusionary. We always lean one way or another when we have to make an important decision, and it is rarely the head that does the leaning. In the elegant phrasing of the seventeenth-century French philosopher Blaise Pascal: "The heart has reasons of which reason knows nothing."[5]

Emotions help us navigate a complex world that we don't fully comprehend. They are our body's way of ensuring that we do what is best for us. Moreover, only the body can carry out the required actions. Minds by themselves are useless: they need bodies to engage with the world. Emotions are at the interface of these three—mind, body, and environment. They are also called *affects*, but since this term has conflicting definitions, I'll stick with *emotions*, defined as follows:

> *An emotion is a temporary state brought about by external stimuli relevant to the organism. It is marked by specific changes in body and mind—brain, hormones, muscles, viscera, heart, alertness, etcetera. Which emotion is being triggered can be inferred by the situation in which the organism finds itself as well as from its behavioral changes and expressions. Instead of a one-on-one relation between an emotion and ensuing behavior, emotions combine individual experience with assessment of the environment to prepare the organism for the optimal response.*[6]

Let's consider the emotion fear. As soon as a monkey spots a snake, it becomes terribly afraid. Similarly, if you step off a curb onto the street and a city bus flies by inches from your face, you will be seized by fear.

Fear makes your body freeze and tremble while your heart rate increases, you breathe faster, your muscles tense, your hair or feathers rise, and you have an adrenaline rush. All this sends oxygen to both brain and muscles to enable you to better cope with your perceived peril. The monkey needs to decide if the snake is dangerous or harmless, and whether its best course of action is to climb a tree, step back, run away, or put up a fight. After your bus encounter, you will check out the traffic and decide if it's safe to cross or else find a zebra crossing. Emotions have the great advantage over instincts that they don't dictate specific behavior. Instincts are rigid and reflex-like, which is not how most animals operate. By contrast, emotions focus the mind and prepare the body while leaving room for experience and judgment. They constitute a flexible response system far and away superior to the instincts. Based on millions of years of evolution, the emotions "know" things about the environment that we as individuals don't always consciously know. This is why the emotions are said to reflect the wisdom of ages.

Getting back to Lisa Parr—she decided to take the chimps' temperature while she was testing them. She taught them to patiently hold out a finger while she put a strap around it and took their skin temperature. In our species, during negative arousal—such as when we see things that upset or frighten us—our skin temperature drops. A fight-or-flight response literally gives us "cold feet" as blood is drawn away from the extremities. In one episode of the *MythBusters* television show, heat sensors were placed on the feet of people who were faced with tarantulas crawling over them or taking a frightening ride in a stunt plane. The drops in temperature were astonishing. Our feet freeze when we're afraid, a reaction that we share with fearful rats, which get cold tails and paws.[7]

Would apes show the same temperature drop? Lisa wondered. First, she played a short video on the screen. It showed a happy scene, such as animal caretakers approaching with buckets full of fruit, or else an unpleasant scene, such as a veterinarian coming with a dart gun—about as close as she could get to a predator. After watching one or the other video, the apes were to choose between two faces on the screen: one with the happy laugh expression of their species, the other with a nervous grin. The goal was to see which face the apes would spontaneously associate with which scene. They had never been trained on these images. On their very first test, they selected the laugh face to go with the happy scene and the distressed grin to go with the unpleasant scene. Seeing the latter video, their skin temperature dropped the way it does in humans and rats facing an unpleasant situation.[8]

I find it hard to explain this outcome without inferring subjective experiences. This is not just about emotions anymore, which may be automatically triggered, but also about *feelings*. Feelings arise when emotions penetrate our consciousness, and we become aware of them. We know that we are angry or in love because we can feel it. We may say we feel it in our "gut," but in fact we detect changes all over our body. How could the apes in Lisa's experiment have selected the right facial expression unless they felt something? Most likely, they felt good or bad from the videos, which then helped them decide which face would go with them. Lisa's temperature measures confirmed that they solved the task emotionally rather than intellectually. Her experiment left us with the intriguing possibility that apes are about as conscious of their feelings as we are.

Most of the time, however, animal feelings remain unknown to us, and all we can do is test their reactions. Experiments have taught us that monkeys and apes are experts at their own facial expressions. They are incredibly quick and accurate at spotting similarities and differences, the way we can instantly tell a smile from a frown. When we showed capuchin monkeys a screen with pictures of different objects—flowers, animals, cars, fruits, human faces, monkey faces—we found that the ones they recognized fastest were the emotional expressions of their own species.[9] These images were a class apart, because expressions are not only meaningful but also engaging. Initially, the monkeys even reacted to them—refusing to touch the picture of a threat face, for example, or lip-smacking at a friendly eyebrow flash. Expressions stir up emotions, or empathy. In fact, it is hard to empathize without a facial connection.

The Swedish psychologist Ulf Dimberg identified the empathic connection in our own species in the 1990s, when he pasted electrodes onto human faces that allowed him to register even the tiniest muscle contractions. He found that people automatically mimic the expressions shown them on a monitor. Most remarkably, they don't even need to know what they're seeing. The pictures of faces can be flashed subliminally (for only a fraction of a second) between pictures of landscapes, and people will still mimic them. They think they're just looking at beautiful scenery, unaware of the faces on the screen, but they feel good or bad afterward depending on whether they were exposed to smiles or frowns. Seeing smiles makes us happy, while seeing frowns makes us angry or sad. Unconsciously, our facial muscles copy these faces, which then feeds back into how we feel.[10]

In real life, then, we can't help but being emotionally affected by others. Our empathic connection with others is like an under-the-table handshake between bodies, perceived as a "vibe," which may be positive and inspiring, or toxic, sapping our energy. It takes time to realize this, because these processes can occur outside our conscious minds. Although Dimberg's research provided wonderful insights in human affairs, he unfortunately ran into enormous resistance and ridicule. For a while, his groundbreaking work remained unpublished because it gave priority to the body, whereas in the West we prefer the mind to be in charge. We like to see ourselves first of all as rational beings, like Darwin drawing up his silly list of pros and cons of marriage. We may camouflage our emotional decisions with rationalizations, saying we need that sports car to beat the traffic, or we need that chocolate because of the antioxidants. For the same reason, science has elevated empathy to a cognitive process. Leaving it a matter of emotions and bodily processes was just not acceptable, so people were said to empathize with others by deliberately putting themselves into their shoes. It was said that we understand others based on a "leap of imagination into someone else's headspace,"[11] or by consciously simulating their situation. The body wasn't part of these theories.

In recent years, however, science has been forced to come around. The body is now front and center to any account of empathy. New brain imaging studies support the involuntary physical process proposed by Dimberg. And research has found that empathy suffers when facial mimicry is blocked, such as when human subjects hold a pencil between their teeth, so that their cheek muscles can't move. Our faces are much more mobile than we think, which helps us connect with others by mimicking their movements. This has become a problem for people whose faces have been injected with Botox. Their muscle relaxation keeps them from mirroring the faces of others, which robs them of feeling what others feel. Botoxed people may look wonderful, but they have trouble empathizing. And the problem is not just in how they relate to others, but also how others relate to them. Botoxed faces look frozen, missing the stream of micro-expressions employed in daily interactions. Their facial unresponsiveness makes others feel cut off, rejected even.[12]

Science's initial skepticism about these bodily processes now strikes us as odd. Who hasn't cried when others cried, laughed when others laughed, or jumped for joy when others jumped? We feel what others feel by making their postures, movements, and expressions our own. Empathy jumps from body to body.

Monkey See, Monkey Do

In 1904 Leo Tolstoy, the Russian novelist, published a children's story that opens with the shocking line: "Wild animals were on show in London. To see them people had to pay money, or bring dogs and cats which were thrown to the wild animals to eat."[13] In the story, a terrified little dog is pushed into the cage of a ferocious lion.

Today crowds would gather outside the gates of this exhibition in angry protest. Attitudes have changed so dramatically that most of us would be horrified, unable to watch. This is telling: I could write a detailed description of a lion attack, and you'd probably read it, but to watch an actual lion's bloody assault on a puppy would be a totally different thing. You'd recoil. The body channel brings events so close to home that we have no escape. We would almost feel as if the lion were attacking us. All we can do is cut off the input by slamming our hands over our eyes. It is hard to imagine, therefore, that previous generations might have enjoyed watching such spectacles. Does this mean we're nowadays more empathic? I'm not sure, because the human capacity for empathy is unlikely to have changed in such a short time. Rather, what has changed is its focus. We regulate empathy by opening or closing a door, depending on who we identify with and feel close to. We open the door wide for friends and relatives, and for animals that we love, but we close it for enemies and for animals we don't care about.

Compared to a century ago, the Western world has been opening its empathy door ever more widely for its favorite pets. They have become part of the family. In 1964 the American president Lyndon B. Johnson, standing before the press on the lawn of the White House, lifted one of his beagles up into the air by its floppy ears. The incident caused an outcry. Huge piles of hate mail arrived at the White House. Afterward Johnson explained that it was a way of making his dog yelp. Well, the dog yelped all right, but the world failed to see the point of this dominance gesture. The stream of protest lasted so long and became so damaging that Johnson was forced to issue a public apology. In fact, he reportedly received more angry mail about this single event than about the entire Vietnam War. Does this mean we care more about the mistreatment of one canine, who survived, than about the violent deaths of over a million human civilians and soldiers? Rationally speaking, I can't imagine that we do, but our visceral reactions are informed by our senses, not by numbers.

A sign of the growing public sensitivity to animal feelings was the national outcry when U.S. president Lyndon
Johnson mistreated one of his beagles. One day in 1964,
he picked up his dog by the ears in front of the press. He
did not lift him off the ground, but pulled him upright.
The dog yelped. This infamous incident, which was photographed, caused a massive outpouring of sympathy for
the animal and condemnation of the president, who was
forced to issue an apology.

Reading about a terrible disaster in a distant land is unlikely to move
us as much as seeing actual pictures or watching interviews with crying
victims. Every charity knows that visuals are essential to obtain donations. Johnson's bad luck was that the dog incident was photographed.
Our greatest sensitivity is always to bodies and faces. Thus, rightly or
wrongly, Anne Frank's portrait has come to stand in for the millions of
Jews killed during the Holocaust. A single tragic picture of a three-year-
old Syrian boy lying facedown on a Mediterranean beach shifted the public debate about a massive refugee crisis that had been going on for years.

We need an individual object of identification, an actual body and face, to open the door to our heart. Michel de Montaigne, a sixteenth-century French philosopher, already knew the power of body language. In grief and sympathy, he said, the role of cognition is grossly overrated by comparison with physical proximity. It's no accident, he said, that we say we are "touched" by an event—using a bodily term—because how we relate to others is greatly aided by actually seeing, feeling, and hearing them.

This body channel is so ancient that we share it with other species. I once saw a chimpanzee, May, unexpectedly give birth at midday. It was right under my office window that overlooked the apes' outdoor area, and May was surrounded by an excited crowd of onlookers. While the chimps elbowed each other aside to get a good look, May stood half upright with her legs spread, then lowered one open hand between them to catch the baby when it popped out. Next to her stood her best friend, Atlanta, an older female—who to my surprise adopted exactly the same posture. Atlanta was not pregnant—she was mimicking May. But she, too, stretched her hand between her legs—her *own* legs. Perhaps there was an element here of modeling, such as "this is what you should do!" the way human parents make chewing movements and slurping noises while feeding their baby with a spoon. Humans and other primates not only mimic others but identify with them so closely that others' situation becomes their own. Finally after a long wait May's baby emerged, and the group stirred. One chimpanzee screamed, and others embraced one another, showing how much they all had been caught up in the emotions of the moment.

Sometimes chimps identify with one another for amusement. Once for a couple of weeks, our juvenile chimpanzees played the fun game of following an injured adult male around. The male didn't do the typical knuckle-walking (leaning his frontal weight on his knuckles) but instead leaned on a bent wrist to protect his bitten fingers. In single file behind him, the juveniles hobbled as pathetically as this unlucky male, looking as if they had been injured as well. Wild chimpanzees in Budongo Forest, in Uganda, were also fascinated with one among them who moved in unusual ways. A nearly fifty-year-old male, Tinka, had severely deformed hands and paralyzed wrists, which meant he couldn't even scratch himself. Tinka developed a scratch technique like the way we stretch a towel between our hands to dry our back. He'd pull a hanging liana plant taut with his foot, then rub his head and body sideways against it. It was an odd procedure—able-bodied chimpanzees would have no use for it. Yet several juveniles thereafter regularly rubbed themselves against lianas pulled down for this purpose, just like Tinka.[14]

As Plutarch said, "You live with a cripple, you will learn to limp." Similar sympathetic locomotion is known of our pets. Within days after a good friend of mine broke his leg, his dog started dragging her own. In both cases, it was the right leg. The dog's limp lasted for weeks but vanished miraculously once my friend's cast came off. This is possible only because dogs, like many mammals, are perfectly in tune with the bodies of others. Not only are they great synchronizers, they enjoy it. Some dogs learn to skip nicely along with children jumping rope, while others will follow a human baby through the house, crawling on their bellies exactly in tandem.

Synchronization and mimicry are common in nature, such as when multiple dolphins jump out of the water as one or when pelicans glide by in seamless formation. We also see it in animals under human care. When two horses are trained to pull a cart together, they will initially push and pull against each other, each following its own rhythm. But after years of working together, they end up acting as one, fearlessly pulling the cart at breakneck speed through water obstacles during cross-country marathons. They will object to even the briefest separation, as if they have become a single organism. The same principle operates among sled dogs. Perhaps the most extreme case is that of a female husky who went blind but still ran along with the rest, based on her ability to smell, hear, and feel them.

Bodily fusion is the core principle. Katy Payne, an American zoologist, worked with African elephants:

Once I saw an elephant mother do a subtle trunk-and-foot dance as she, without advancing, watched her son chase a fleeing wildebeest. I have danced like that myself while watching my children's performances—and one of my children, I can't resist telling you, is a circus acrobat.[15]

A century ago Theodor Lipps, the German psychologist who inspired the term *empathy*, explained *Einfühlung* (German for "feeling into") with a strikingly similar example: the case of a high-wire artist. While we are watching the artist's performance, we emotionally enter his body and share his experience as if we were on the rope with him. Lipps was the first to recognize this special channel we have to others. We can't feel anything that happens outside ourselves, but by unconsciously becoming one with another's body, we gain similar experiences, feeling their situation as if it's our own.

That explains why our reactions are instantaneous. Imagine the circus acrobat falls, and the audience empathizes based only on a mental re-creation. Such a process takes time and effort, so my guess is that they wouldn't react until the acrobat's broken body was lying in a puddle of blood on the floor. But that's not what happens. The audience's reaction is instantaneous: hundreds of spectators utter "ooh" and "aah" at the *very instant* the acrobat's foot slips. Acrobats sometimes slip on purpose, with no intention of falling, precisely because they know their audience is with them every step of the way. I sometimes wonder where Cirque du Soleil would be without this kind of empathic connection.

About twenty-five years ago the body channel received a tremendous boost from the discovery of mirror neurons in a laboratory in Parma, Italy. These neurons are activated when we perform an action, such as reaching for a cup, but also when we see someone else reach for a cup. These neurons don't distinguish between our own behavior and that of someone else, so they allow one individual to get under another's skin. Their actions become our own. This discovery has been hailed as being of equal importance to psychology as the discovery of DNA was for biology, because of its profound implications for imitation and other forms of bodily fusion. It explains why words come automatically to our mouths when we watch the stammering King George VI in the 2010 film *The King's Speech,* and why Atlanta copied May's posture and movements.

Amid all the brouhaha surrounding mirror neurons, however, we should not forget that they were discovered not in humans but in macaques. And still today, the evidence for "monkey-see, monkey-do" neurons in other primates is better and more detailed than for their equivalent in the human brain. Mirror neurons probably help primates imitate others, such as when they open a box the way a trained model does, when they synchronize with each other while pressing buttons, or when in the wild they remove the seeds from a fruit the way their mother does.[16] Monkeys in different groups process fruits in slightly different ways, and the young faithfully copy their elders.[17] Primates are in fact natural conformists. Not only do they imitate, they also like to be imitated. In one experiment, two investigators gave a capuchin monkey a plastic ball to play with. One investigator mimicked every move the monkey made with the ball—throwing it, sitting on it, banging it against the wall—while the other did not. By the end, the monkey clearly preferred the one who had imitated him.[18] In a similar study, human adolescents going out with a date were instructed to mimic their date's every move, such as picking up a glass, leaning an elbow on the table, or scratching their head. The date reported liking him or her much better than those whose date acted independently. They don't realize why they feel differently, but evidently on some level we regard imitation as a compliment.

It is easy to see how this works when someone we are close to yawns in our presence. It is almost impossible not to yawn along with them. I have attended lectures on yawning (using fancy terms such as *pandiculation*), in which the whole audience sat with mouths open most of the time. Yawn contagion relates to empathy, because the humans most prone to it are also the most empathic on other measures, and women—who on average score higher on empathy than men—are more sensitive to the yawns of others. On the other hand, children with empathy deficits, such as those with Autism Spectrum Disorder, often lack yawn contagion. This knowledge has led to quite a few studies to see how and when we catch the yawns of others, and whether other animals do the same. We now know that dogs and horses yawn in response to human yawns—dogs do so even if they only *hear* their owner yawning—and that yawns often spread among monkeys in a group.

We taught our chimps to press one eye to a hole in a bucket to watch an iPod held up at the other end. This way we could test their private reaction to videos of yawning apes. As soon as they saw the yawns, they started yawning like crazy themselves. They only did so, however, if they personally knew the apes in the videos. Videos of strangers left them cold. So it was not just a matter of seeing a mouth open and close—they needed to identify with the yawning ape in the video.[19] The same role of familiarity is known among humans. An undercover field study in restaurants, waiting rooms, and train stations found that if a man stands next to his wife, who yawns, he will yawn along with her. But if he stands next to a stranger who yawns, he will remain unaffected. Empathic reactions are always stronger the more we have in common with the other and the closer we feel to them.[20]

Let me finish Tolstoy's story of the lion and the puppy. Meeting the big cat, the poor puppy quickly rolled onto his back while frantically wagging his tail. This act of surrender must have appeased the lion, because he refrained from pouncing. Not only that, the two became best friends. While this may seem implausible, there are enough contemporary examples of odd animal friendships—elephant and dog, owl and cat, even lion and dachshund—to not dismiss Tolstoy's story out of hand. It always boils down to how bodies interact, such as how full the lion's belly was and how convincing the dog's rollover.

Kissing the Sore Spot

When the body channel helps spread emotions from one individual to another, it's not just about yawning anymore, or mimicking, but about feeling what others feel. Even though still rooted in bodily connections, here we're getting close to real empathy. *Emotional contagion,* as it is known, begins at birth, such as when one baby cries upon hearing another baby cry. On airplanes and in maternity wards, babies sometimes chorus like frogs. You might think they cry in reaction to any kind of noise, but studies have shown they respond specifically to the cries of same-age babies. Girl babies do so more than boy babies. That the emotional glue of society emerges so early in life reveals its biological nature. It is a capacity we share with all mammals.

In real life, a wild female orangutan will be skillfully swinging from one tall tree to the next. Her young offspring, trying to follow her through the tree canopy, comes to a stop: the gap between the next two trees is too wide for him. He whimpers and desperately calls for her help. Hearing him, she may whimper herself and hurries back to make a bridge for the juvenile. She grabs a branch from one tree with one hand and a branch from another tree with her other hand or her foot, then pulls the two trees closer to each other while draping herself between them, enabling her offspring to cross over by using her body as a live bridge. This everyday sequence is driven by emotional contagion—the mother being distressed by her offspring's whimpers—combined with intelligence, which allows the mother to understand the problem and come up with a solution.

What is most astonishing is the pull of negative emotions. You'd think that signals of fear and distress would be highly aversive, but a recent study found that mice are actually drawn to other mice in pain.[21] I am quite familiar with this phenomenon in young rhesus monkeys. Once an infant accidentally landed on a dominant female, who bit him. He screamed so incessantly that he was soon surrounded by other infants.

I counted eight of them in the baby pile, all climbing on top of the poor victim, pushing, pulling, and shoving each other aside. That obviously did little to alleviate the first infant's fright. But the monkeys' response seemed automatic, as if they were just as distraught as the victim and sought to comfort themselves as much as the other.

This can't be the whole story, though. If these baby monkeys were trying to calm themselves, why would they need to get close to the victim rather than run to their mothers? In fact, they sought out the actual source of distress rather than a guaranteed source of comfort. Baby monkeys do this all the time without any indication that they know what is going on. They seem attracted to the distress of others like moths to a flame.

We like to read concern into such behavior, but they probably don't grasp what happened to the first infant. I call this kind of blind attraction to those in trouble *pre-concern*. It is as if nature has endowed children and many animals with a simple rule: "If you feel another's pain, get over there and make contact!" It is good to realize, however, that any theory of strict self-preservation would predict the exact opposite. If others around you are screaming and whimpering, there's a good chance they're in danger, so the wise thing would be to remove yourself. The same applies to sounds of distress. If high-pitched screams grate on your ear, the logical thing to do would be to cover your ears or move away. But many animals do the opposite—they get closer to find out what's going on, even when the sounds of pain are barely audible. It's all about the emotional state of the other. That mice, monkeys, and many other animals actively seek out those in trouble doesn't fit purely selfish scenarios and proves the fundamental flaw of the sociobiological theories popular in the 1970s and '80s.

In sociobiological depictions of nature as a dog-eat-dog place, all behavior boiled down to selfish genes, and self-serving tendencies were invariably attributed to "the law of the strongest." Genuine kindness was out of the question, because no organism would be so stupid as to ignore danger in order to assist another. If such behavior did occur, it must be either a mirage or a product of "misfiring" genes. The infamous summary line of this era, "Scratch an altruist, and watch a hypocrite bleed,"[22] was quoted over and over with a certain amount of glee: altruism, it said, must be a sham. The line was used to rebuff bleeding-heart romantics and wishful thinkers, who naively believed in human goodness. Not coincidentally, this was also the time of Ronald Reagan and Margaret Thatcher, as well as Gordon Gekko, a fictional character in the 1987 film *Wall Street*: Gekko believed that greed was what made the world go around. Nearly everyone was running after a simple idea that was plainly at odds with the way social animals, including humans, have been shaped by natural selection.

Fortunately, we don't hear much about "selfish genes" anymore. Buried by a mass of fresh data, the idea that behavior is invariably self-serving has died an inglorious death. Science has confirmed that cooperation is our species's first and foremost inclination, at least with members of the in-group, so much so that a 2011 book about human behavior by Martin Nowak was entitled *SuperCooperators: Altruism, Evolution, and Why We Need Each Other to Succeed*. When people in a neuroimaging experiment were given a choice between a selfish and an altruistic option, most opted for the latter. They went with the selfish choice only if there were good reasons to avoid cooperation.[23] Many studies support this view, saying that we tend to be kind and open to others unless something holds us back. I sometimes joke that this must be why Ayn Rand, the Russian-American novelist and would-be philosopher, needed such boring heavy tomes full of bloodless characters to make her case. Her main point was that we are unalloyed individualists, but she had to work hard to convince us, because deep down everyone knows that this is not who or what we are. Rather than a description of our species, Rand offered a counterintuitive ideological construct.

The default mode of the human primate is intensely social, as reflected in our favorite activities, from attending sports matches and singing in choirs to partying and socializing. Given that we derive from a long line of group-living animals, which survived by helping one another, these tendencies are entirely logical. Going it alone has never worked out for us.

Nadia Ladygina-Kohts provided a typical example of the prosocial nature of our primate kin, including the pull of distress signals, in her adopted chimp, Joni:

> *If I pretend to be crying, close my eyes and weep, Joni immediately stops his play or any other activities, quickly runs over to me, all excited and shagged, from the most remote places in the house, such as the roof or the ceiling of his cage, from where I could not drive him down despite my persistent calls and entreaties. He hastily runs around me, as if looking for the offender; looking at my face, he tenderly takes my chin in his palm, lightly touches my face with his finger, as though trying to understand what is happening, and turns around, clenching his toes into firm fists.*[24]

What better proof of simian sympathy than the fact that an ape who refused to descend from the roof for food did so instantly upon seeing his mistress in misery? When Kohts pretended to cry, Joni would look into her eyes, and "the more sorrowful and disconsolate my crying, the warmer his sympathy." When she slapped her hands over her eyes, he tried to pull them away, extending his lips toward her face, looking attentively, slightly groaning and whimpering.

When animals and children begin to understand what is the matter with a person in distress, they leave behind blind attraction and demonstrate *empathic concern*. They try to ease the pain, the way Joni did with Kohts. It is also the way human parents react when their children scrape a knee, bump a head, or get punched or bitten by another child. The quickest way to get them to stop crying is to kiss the sore spot.

The early development of this behavior has been studied in our species by filming children in their homes. The investigator asks an adult family member to pretend to cry or act as if they are in pain, in order to see what the children do. In the film, the children look worried while approaching the distressed adult. They gently touch, stroke, hug, or kiss the adult. Girls do so more than boys. The most important finding was that these responses emerge early in life, before the age of two. That toddlers already express empathy suggests it is spontaneous, because it is unlikely that anyone has been instructing them how to respond.[25]

For me, the real eye-opener was that the children behaved exactly like apes. Apes not only approach a distressed other but go through the same routine of touching, hugging, and kissing. After watching films of the human study, I realized on the spot that all along I had been studying empathic concern, because why should I adopt a different terminology? Many animals, from dogs to rodents, and from dolphins to elephants, exhibit comforting behavior, even though each species uses its own gestures. In fact, in the same homes where the children were filmed, the psychologists accidentally discovered that dogs responded to the distressed person as well, putting their heads in their lap or licking their face. More targeted studies subsequently confirmed this behavior.[26]

Sure enough, not everyone liked to hear dogs and apes described as being empathic, but over the years resistance has diminished. The idea of animal empathy is now fairly well established. After all, no one is claiming that dogs have all the mental capacities that humans bring to the job of understanding others. Many different levels mark empathy. But we can certainly recognize in dogs sensitivity to the emotions of others, adoption of similar emotions, and expressions of concern. This is in fact why we consider the dog man's best friend. In primates, empathy is so obvious and common that by now dozens of studies have examined "consolation," the tendency to comfort and reassure those who have gone through a painful experience. To document how primates console, we simply wait for a spontaneous incident that stresses them out—a fight, a fall, a frustration—then observe how others console them. Consolation through body contact has a calming effect and is typical of close social relationships. It is also very effective. One minute an ape is screaming her lungs out and slapping herself with spasmodic arm movements, hitting her sides in a noisy tantrum, due to her failure to get the food she was begging for. The next moment, while a friend keeps her tight in a hug, her screams taper off into soft whimpers.[27]

Since consolation behavior is not at all limited to bonobos and chimpanzees, I was happy when one day a student joined my team who wanted to study elephants. Together with Josh Plotnik, we observed the largest land mammal, well known for its social bonding and mutual assistance. At an open-air sanctuary in northern Thailand, where rescued Asian elephants roam in semiliberty, a cow named Mae Perm would rush to the side of her friend, a blind cow named Jokia, whenever the latter needed her—Mae Perm acted as Jokia's "seeing eye dog." Those two were always vocally in touch, trumpeting and rumbling at each other. If Jokia was upset or spooked by anything, such as the roar of a bull elephant or a distant traffic noise, both elephants would spread their ears and raise their tails. Mae Perm might produce reassuring chirps and caress Jokia with her trunk or place it into Jokia's mouth. This made her immensely vulnerable (nothing is more sensitive and important to an elephant than the tip of her trunk) but validated her trust in the other. Jokia would do the same by placing her trunk into Mae Perm's mouth, so the trust was mutual.

If other elephants were around, they might react just as agitated as Jokia did, raising their tails, flaring their ears, and sometimes urinating and defecating while chirping. They'd position themselves in a protective ring around her.

Josh found ample evidence for emotional contagion and consolation in these pachyderms.[28] Many people consider its existence so self-evident, though, that he sometimes gets asked why his studies were even needed. Doesn't everyone know that elephants have empathy? In a way, I'm thrilled to hear this question, because it shows how well established the idea of animal empathy has become. Science progresses amid enormous skepticism, though, and anyone who remembers the fierce resistance to this idea, as I certainly do, realizes that without solid data, it would never have taken hold. But it clearly has, in the same way that we now accept that the heart pumps blood and that the earth is round. We can't even imagine that people used to think otherwise.

Yet even after reaching this point with regard to the emotional sensitivity of mammals, we still need studies to learn how it works and under which circumstances it finds expression, because empathy is never the only option. Mae Perm, for example, was not beyond taking advantage of Jokia's condition to filch her food.

To grasp another's impairment also gives you ways to exploit it.

The Good and the Bad

Paradoxically, the reason humans can be so unfathomably cruel to each other relates to empathy. The typical definition of empathy—sensitivity to another's emotions, understanding another's situation—says nothing about being nice. Like intelligence or physical strength, it is a neutral capacity. It can be used for good or evil, depending on one's intentions. Being an effective torturer, for example, requires knowing what hurts the most. A used-car salesman may empathize and joke with you, only to sell you a crappy car for too much money. Despite the rosy-colored assumptions surrounding the term, empathy is an all-purpose capacity.

It is true, though, that most of the time, empathy favors positive outcomes. It evolved in order to assist others, initially in parental care, the prototypical form of altruism and the blueprint for all other kinds. In mammals, mothers are obliged to care for offspring, while for fathers it is optional. Mammals need to nurse their young, and only one sex is equipped to do that. Not surprisingly, therefore, females are more nurturing and empathic than males. Consolation behavior is more typical of female apes than males, and the same is true for our species. A recent analysis of surveillance camera footage of store robberies confirmed that victims of these unsettling incidents received physical consolation from women far more often than from men.[29] This sex difference holds for all mammals studied thus far, and in our own species empathy differences are reflected even in scholarship and science. Numerous men have written about the "puzzle" of altruism, as if it were a perplexing thing that comes out of nowhere and needs special attention. They regard altruism as such a tough cookie to account for, so counterintuitive, that we have libraries full of learned speculations about why and how it may have evolved. This literature overlooks maternal care because it is not nearly baffling enough. Given how easily behavior in favor of one's own progeny is explained, why dwell on it?

In contrast, I don't know of a single woman scientist who has got carried away by the puzzle of altruism. Women would find it hard to leave out maternal nurturance and the constant worry and attention that it entails. Writing on cooperation, Sarah Hrdy, an American anthropologist, proposed an "it takes a village" theory, according to which the human team spirit started with collective care for the young.[30] Similarly, Patricia Churchland, an American philosopher well versed in neuroscience, treats human morality as an outgrowth of offspring care tendencies. The female body co-opts the neural circuitry that regulates its own functions to include the needs of the young, treating them almost like an extra limb. Since our children are part of us neurologically, we protect and nurse them unthinkingly, the way we do our own bodies. The same brain circuitry provides the basis for other kinds of nurturance as well, including care aimed at distant relatives, spouses, and friends.[31]

This maternal origin explains the pervasive sex difference in empathy, which starts early in life. At birth, girl babies look longer at faces than do boy babies, who look longer at mechanical toys. Later in life, girls are more prosocial than boys, better readers of facial expressions, more attuned to voices, more remorseful after having hurt someone, and better at taking another's perspective.[32] The same differences have been found in self-report studies of human adults. We also know that empathy is enhanced by oxytocin sprayed into the nostrils of both men and women, thus fooling them with the maternal hormone par excellence. As a result, we barely notice our own daily efforts on behalf of our progeny and even joke about the arm and leg that they cost. Distant relatives and friends recruit less help, but the underlying satisfaction remains the same. Adam Smith, the eighteenth-century Scottish philosopher, understood as no other that the pursuit of self-interest needs to be tempered by "fellow feeling." He said so in *The Theory of Moral Sentiments* (1759), a book not nearly as popular as his later *The Wealth of Nations* (1776), which founded the discipline of economics. He famously opened his first book with:

> *How selfish soever man may be supposed, there are evidently some principles in his nature, which interest him in the fortune of others, and render their happiness necessary to him, though he derives nothing from it except the pleasure of seeing it.*[33]

In order to survive, we need to eat, make love, and nurse. Nature has made all these activities pleasurable, so we engage in them easily and voluntarily. Nature has done the same for empathy and mutual help, by making us feel good when we do good—the "warm glow" effect of altruism. Altruism activates one of the most ancient and essential mammalian brain circuits, helping us care for those close to us while building the cooperative societies on which our survival depends. By seeking the origin of human altruism in its oldest and most convincing expression, we take the puzzle right out of it.

The neural mechanisms behind animal empathy are less known, because it is impossible to conduct similar studies on apes, elephants, dolphins, and so on. They don't fit into an ordinary brain scanner, or they won't sit still so they can be tested while awake. Rodents, on the other hand, are used in neuroscience all the time. Here at Emory University, where I work, James Burkett discovered that prairie voles console each other during stress. In this tiny rodent, males and females are attached to each other in what is known as a monogamous pair-bond and raise their pups together. If one mate is upset by anything, the other is affected to an equal degree. This is true even if the other was not present during the stressful event. Afterward the level of corticosterone—a stress hormone—in the male's blood perfectly matches his mate's, and vice versa, indicating a strong emotional linkage. James further found that mates groom each other more if one of them is stressed and that this activity calms them down. If the voles are made immune to the effects of oxytocin, on the other hand, they don't respond to the other's stress, suggesting that oxytocin is critical. This makes vole empathy fundamentally similar to human empathy, also in the brain.[34]

Human emotional contagion has been tested the way the voles' was, by measuring stress hormones. Since the average person fears public speaking more than death, subjects in a study were asked to address an audience. Afterward all participants were invited to spit into a cup, which allowed scientists to extract a hormone associated with anxiety. They found that with confident speakers, the audience followed every word, feeling relaxed, but with nervous ones, the speaker's discomfort rubbed off on the audience. The hormone levels of speakers and audiences converged the way they did between vole mates.[35] These similarities hint at what biologists call *homology*, or traits derived from a common ancestor. In the same way that our hands are homologous to primate hands, mammalian empathy is homologous across species in that it works in the same way and has a common evolutionary origin.

In the long-ago time of Adam Smith, before we had the term *empathy*, all this fell under *sympathy*. Nowadays, however, sympathy means something else. Empathy seeks information about another and helps us understand their situation, whereas sympathy reflects actual concern about the other and a desire to improve their situation.[36] My own profession as a primate watcher, for example, depends heavily on empathy but not on sympathy. It would be terribly boring to observe animals for hours on end without identifying with them, without feeling any ups and downs associated with their ups and downs. The sudden death of a companion, the birth of a healthy infant, the joy of receiving a favorite food—all this rubs off on the human observer. Scientists often declare that objectivity is their goal, but I beg to differ: all that has given us is a cold, mechanistic view of animals. The science may be objective, but it completely misses out on animal emotions. Some of the greatest pioneers in the study of animal behavior rejected this approach by stressing the need to identify with and get close to our subjects. Kinji Imanishi, the founder of Japanese primatology, and Konrad Lorenz both proposed empathy as a gateway to the animal mind. Lorenz went so far as to say that anyone who has lived with a dog and is *unconvinced* that dogs have feelings like us is psychologically deranged, dangerous even.[37]

I consider empathy my bread and butter, as I have made many a discovery by getting under the skin of my subjects. This is not the same as sympathy, of which I have plenty as well, but it is less spontaneous than empathy, more subject to calculation. Some people show nearly unlimited sympathy for animals, such as those who rescue stray pets and nurse them back to health. Abraham Lincoln apparently interrupted a journey and soiled his good pants to pull a squealing pig from the mud. Sympathy is action-oriented. It is often rooted in empathy but goes beyond it.

Whereas sympathy is by definition positive, empathy doesn't need to be, especially if the capacity to understand others is turned against them. Small-brained animals, such as sharks and snakes, are probably incapable of doing so intentionally. These animals have excellent abilities to hurt and damage others, but without the slightest idea of their impact. Most "cruelty" in nature is of this kind: cruel in outcome, but not on purpose. The brains of apes, on the other hand, are sufficiently complex to knowingly inflict pain. They can recruit their capacity to understand others in order to torment them. Like boys throwing rocks at ducks in a pond, apes sometimes harm others just for fun. In one game, juvenile lab chimpanzees enticed chickens behind a fence with breadcrumbs. Every time a gullible chicken approached, the chimps hit it with a stick or poked it with a sharp piece of wire. They invented this "Tantalus" game, which the chickens were stupid enough to play (although it was no game to them), to fight boredom. They refined it to the point that one ape would be the baiter, another the hit man.

In our own studies, we saw something related, albeit less cruel. We were interested in the hooting and food grunts that chimps use to announce a food bonanza, so we set up a test in which some of our chimps discovered a box full of apples inside a building with a small open window nearby. All their friends outside could look in and see what was going on. They would gather at the window, pushing each other aside while reaching in with their arms, begging with open hands for apples. Occasionally, adult food possessors would actually hand them a few fruits, even though they could easily have kept them for themselves. Juveniles, by contrast, saw this as the perfect occasion to tease those outside. Sitting a short distance from the window, they'd hold up a shiny red apple for everyone to see, only to quickly pull back as soon as someone reached for it. The rich kids were taunting the poor.[38]

In nature, chimpanzees have been observed tormenting small animals such as squirrels or hyraxes. They seem to derive pleasure from it, because they laugh while doing so, as if it's fun. Koichiro Zamma, a Japanese fieldworker, described how Nkombo, an adult female in the Mahale Mountains National Park in Tanzania, dragged and swung a squirrel around for six minutes until the animal gave its last desperate cry and died. "It looked like a bullfight," Zamma wrote: "Nkombo as a matador waved a red cloth (her lower arm) in front of a bull (a squirrel), and stabbed (bit) it. This movement seemed like a kind of social play having the characteristic of teasing because Nkombo allowed the squirrel to counterattack and Nkombo showed a play face," or laugh expression.[39] After the squirrel died, Nkombo's behavior changed dramatically. She ceased provoking it and held it by its body, not by its tail as before, which to Zamma suggested that Nkombo understood the change in the animal's condition. She abandoned the body without eating from it.

The possibility that a species other than our own is not only empathic but can be deliberately mean gives extra weight to killings observed in the wild. Like lots of males in the animal kingdom, male chimpanzees fight over territory, but they also occasionally go out of their way to deliberately finish off a rival. Several may set out on a marching patrol around the territory, stalk a victim across the border in total silence to surprise him in a fruit tree, then horrifically bite and hit their enemy to a pulp, to the point of incapacitating him, after which they leave him for dead. I have witnessed similar violence in captivity, once even including emasculation, which at the time was speculated to have been an accident or an artifact of the living conditions. But it is now well established that wild male chimps do the same. In fact, the gruesome attack I saw seems fairly normal for the species. Instead of regarding death and castration as unfortunate by-products of male-to-male combat, I now tend to view both as intentional. Given that these primates are capable of concern for others based on an appreciation of their situation, why not also assume that they can kill in order to kill, hence are capable of murder?

When critics bring up this kind of savagery to discredit the idea of empathy in chimpanzees ("You know those guys kill each other, right?"), I draw attention to our own fine species. No one ever argues against the human capacity for empathy on the grounds that under some circumstances people kill. Our attitudes vary with the situation, giving us the honor of being both the kindest and the nastiest animal on earth. I don't see much of a contradiction, though, because concern and cruelty have more in common than we think. They are two sides of the same coin.

In the third century, Tertullian of Carthage, an early Christian theologian, had a most unusual vision of heaven. While hell was a place of torture, heaven was a balcony from which the saved ones could watch hell, thus enjoying the spectacle of doomed souls frying in the fire. What an odd idea! For many of us, it's almost harder to watch the suffering of others than to suffer ourselves. Tertullian's balcony strikes me as about as unpleasant as hell itself.

But what about our rivals—do we feel the same about them? A German neuroscientist, Tania Singer, who explored this issue discovered yet another intriguing sex difference. When people having their brain scanned watched the hand of another person being mildly shocked, the pain areas in their own brain lit up, showing that they shared the other's pain. This is typical of empathy. But it happened only if the partner was someone likable, with whom the subject had played a friendly game before the scanning session. On the other hand, if the partner had played unfairly against them before the session, the subjects felt cheated, and seeing the other in pain had less of an effect. The door to empathy had shut. For the women, it was still partially open—they still showed mild empathy. But for the men it closed completely—in fact, seeing the unfair player getting shocked activated the *pleasure* centers in men's brains. They had moved from empathy to justice and welcomed the punishment of the other. Their main sentiment was Schadenfreude.[40]

If there is a Tertullian heaven, it must be men watching their enemies burn.

Rat Sympathy

My favorite story of human sympathy remains the Parable of the Good Samaritan. It begins when a priest and a Levite callously pass a victim by the side of the road without even stopping. They are extensively familiar with all the texts urging us to love our neighbor, but they clearly have different priorities. Only the Samaritan, a religious outcast, feels compassion and provides assistance. The message of the story is to be wary of ethics by the book instead of by the heart. It is a wonderful message to keep in mind when scholars or politicians pooh-pooh tender feelings as something we can easily do without. Who needs fellow-feeling? The psychologist Paul Bloom wrote a whole book entitled *Against Empathy* arguing that we are rational beings, hence should ground our morality in logic and reason. If we think hard enough, preferably guided by science, we'll end up with perfectly thought-out choices between right and wrong. What could be better than objective morality?

His position is positively terrifying in the light of recent history. Science and reason, lacking a humane anchor, can be used to justify basically anything, including abhorrent practices. They have given us sound economic arguments for slavery as well as medical justifications for the use of prisoners as guinea pigs. They have urged us to improve the human race through forced sterilization and genocide. Not so long ago, eugenics was a perfectly respectable science, taught at universities all over the world. To winnow out inferior races made sense to those who viewed themselves as superior. This is what you get when logic rules and the heart is left out of the equation. We learned the consequences of this line of rational thought during World War II, when we also learned that the greatest heroes are not those who think like everyone else but whose empathy for others makes them disobey ghastly orders. They secretly feed starving prisoners or hide members of persecuted groups in basements and attics. A Polish nurse, Irena Sendler, smuggled hundreds of Jewish children one by one out of the Warsaw ghetto. She did so not based on some lofty moral principle but out of natural empathy.

But many rationalists view empathy and sympathy as weaknesses, regarding them as too impulsive and unruly. But isn't this precisely their strength? Empathy feeds our interest in others. The pleasure we take in the company and well-being of others is part of our biology. It's who we are, hence requires no moral justification. We also don't need the Bible for illustrations, because reports of surprising acts of kindness reach us every day. People jump into icy rivers to save strangers, drag them off the track of an oncoming subway train, or shield their bodies during a shooting. They make these sacrifices without giving much thought to the consequences, which is why heroes often seem bewildered by the attention they receive. To their minds, they just did what needed to be done. Hardly a day goes by without a new Internet video featuring a dog dragging a wounded companion off a highway, an elephant preventing a calf from being swept away by a river, or humpback whales rescuing a seal from marauding killer whales. Most of these rescues occur in reaction to signs of distress. It is the prototypical mammalian helping response on behalf of offspring in danger, but extended to others, sometimes even different species.

Even more intriguing is aid in the absence of any clear distress signal. Here helpers appreciate, just from taking in the situation, what kind of action is called for. As an illustration, let me return to the bonobo enclosure at the San Diego Zoo, when it still had a wet moat. One day the keepers had drained the moat for scrubbing and were ready to fill it up again. They went to the kitchen to turn on the valve, but all of a sudden the group's alpha male, Kakowet, appeared in front of the kitchen window screaming and waving his arms. The keepers said it was almost as if he were talking.

As it turned out, several young bonobos had jumped into the dry moat but were unable to get out. If the water flow had gone as scheduled, they would have drowned, because apes don't swim. The keepers provided a ladder, and with human assistance all the bonobos got out except for the smallest one, who was pulled up by Kakowet himself. Kakowet's frantic intervention indicates that he realized how the water supply was controlled and who controlled it, and that filling the moat would have disastrous consequences. He took action before any emergency could arise.

Individuals sometimes bring water or food to aging group mates. In our chimpanzee colony, Peony, an old female with arthritis, on some days could barely walk, not even to the faucet. Younger females would go there, suck up a mouthful of water, and bring it to Peony. She'd open her mouth wide, into which they would spit a jet of water. On other occasions, a younger female would help Peony join a cluster of grooming chimps in the climbing frame, by placing both hands on her ample behind and pushing her up. In the wild, an old female who had lost the ability to climb trees was brought fruits by her daughter, who descended with hands full of them.

At ChimpHaven, a sanctuary in Louisiana that I support and work with, the chimpanzees dwell on huge forested islands. They have "retired" from research laboratories, which means that they are often naive about grass, trees, and the outdoors. Chimps with experience teach those without. Once a female named Sara saved her close friend, Sheila, from a venomous snake. Sara saw the snake first and sounded the alarm by *waa*-barking loudly, so that everyone knew about it. But when Sheila ran over to take a look, Sara had to hold her back by her arm, vigorously dragging her away. While Sara poked the snake with a stick to test it out, she continued to hold Sheila back. She must have assumed Sheila wanted to grab the serpent, which would have been a fatal mistake.

I could offer dozens more examples for primates, and at least as many for dolphins, canines, birds, and so on. Elephants, especially, offer rich material, such as the way they rescue a calf from the deadly suction of a mud hole by stepping into it and putting their trunks underneath the struggling little one to lift it up. A viral video from a South Korean zoo showed a calf slipping into a pool, whereupon the mother panics at the edge. The calf's auntie hurries over and forcefully nudges Mom with her head toward steps that lead into the pool. Both cows then enter the water together, swim toward the calf, and herd it toward the same steps to climb out. Since the calf is an excellent swimmer, using its trunk as a snorkel, the adults' panic seems a bit overblown, which is why expert Joyce Poole commented on the incident, "Elephants are drama queens." For me, the most intriguing part was that the auntie knew how to get the calf out of the pool but urged Mom to take the lead.

Lots of species seem to grasp the needs of others and spontaneously act on their behalf. But instead of offering more stories, I'd like to focus here on a few experiments, because they are the only way to nail the evidence down. Observations are too open-ended for drawing firm conclusions. Experiments offer controlled situations in which animals have various options while ruling out potential self-interest. Until recently, experiments on helping behavior hardly existed, though, because scientists widely assumed that only humans care about the well-being of others. Animals were said to be indifferent to others' fate. Sometimes scientists stated this most dramatically, stressing the nobility of human nature or claiming that a relatively recent evolutionary "spark" made our ancestors different. Much as the Church Fathers refused to take a peek through Galileo's telescope convinced that there was nothing to see, scientists had low expectations that impoverished animal behavior research for most of last century. Why test animals for faculties they can't possibly possess? Things have begun to change, however. Since everything humans do must have antecedents or parallels in other species, including aiding behavior, the latter has now become a respectable topic of study.

One remarkable series of tests, by the American anthropologist Brian Hare and his colleagues, concerned the most empathic ape, the bonobo.[41] Bonobos are as close to us as chimpanzees but are considerably more sensitive and gentle. Their use of sexual contact to defuse tensions prompted me long ago to dub them the "Make Love—Not War" primates, a label that has stuck. In Hare's creative experiments, they lived up to their reputation. In one test, the researchers gave a young bonobo a whole pile of fruits that he could eat all by himself. If left alone, he'd consume it in its entirety. But often the ape could see a companion sitting behind a mesh door, which he knew how to unlock. The first thing many bonobos did, before consuming the fruits, was to open that door and let the other one in. This move cost them half their goodies, because now they had to share. If there was no one behind the door, on the other hand, they rarely touched it. Even more striking were tests in which the apes were given a chance to produce food for others without getting anything themselves. They could pull a rope that would open a door to give another bonobo access to fruits, but they themselves couldn't join in the feast. They still pulled the rope, however, even though—to recall Smith's words on sympathy—they "derived nothing from it except the pleasure of seeing it."

This kind of test is not just about altruism, which can come about in various ways, but about *prosocial* tendencies, defined as the intention to make life better for others. One of my team members, Vicky Horner, explored prosocial versus selfish choices in chimpanzees under controlled conditions. Vicky would call two chimps into the Cognition Building and place them side by side with mesh between them. Our first test involved old Peony and Rita, an unrelated female. Peony received a bucket full of colored plastic tokens, half of them green and the other half red. She had learned to select one token at a time and hand it to us but knew nothing about the two colors. Whichever color she picked, she'd always receive a reward for it. The only difference was what Rita would get. Red tokens were "selfish" in that they rewarded only Peony, whereas green tokens were "prosocial" in that they rewarded both partners. Choosing many times in a row, Peony began to prefer green tokens two out of three times. Other chimpanzee pairs made up to nine out of ten prosocial choices. If we tested a chimp alone, on the other hand, the colors made no difference to them. The prosocial preference developed only if a partner gained from it.[42]

But the glass-is-half-full-versus-half-empty debate remains: while we were mightily impressed by our apes' prosociality, critics pointed out that they failed to be prosocial all the time. Chimps must be "mean-spirited" creatures, they said, because why else would they deliberately withhold rewards from a partner? This attempt to reclaim some of the lost ground for the idea that only humans care about others failed, however. Chimps are complex beings, who vary their behavior all the time. I don't know of a single task that they perform one hundred percent, even if they know perfectly well how it works. Humans are no different: our performance also varies with circumstances, moods, attention, and partners. Reading up on human prosocial choice, we found the exact same variability as among chimps. Seven-to-eight-year-old children, for example, are prosocial only three-quarters of the time, meaning they make selfish choices one-quarter of the time. Other studies suggest the same. Like chimps, humans are never perfectly prosocial.[43]

In Japan, Shinya Yamamoto conducted a test in which chimps could help each other, but only if they'd adopt the other's perspective. His results were similar to the anecdotes recounted above about chimps understanding when others needed food or water, or were about to make a stupid move on a snake. Yamamoto brought this kind of insightful assistance under experimental control. He gave an ape two ways to obtain orange juice: she could use a rake to bring a container close, or she could use a straw to suck juice through. But she didn't have either a rake or a straw available. Next to her, in a separate area, sat another chimp who had a whole set of tools. After taking one look at her problem, he'd pick out the right tool for the task and hand it to her through a small window. If he had been unable to see her situation, however, he'd pick tools at random, indicating that he had no idea what she needed. So chimps not only readily assist others but are capable of taking their specific needs into account.[44]

We still know little about these capacities, but clearly apes are not nearly as selfish as assumed, and they might actually beat the average priest or Levite when it comes to humane behavior. For both practical and ethical reasons, however, we have virtually no experiments about costly forms of altruism, such as when individuals risk their lives to help others. No scientist is going to deliberately throw a chimp into a river to test whether another will save him, but we do know from actual observations that they will. Zoos often accommodate apes on islands surrounded by wet moats, and we have reports of them trying to save companions who have fallen in, sometimes with a fatal outcome for both. One male lost his life when he waded into water to reach an infant, who had been dropped by an inept mother. At another zoo, an infant chimpanzee hit an electric wire and panicked, jumping off his mother into the water, whereupon mother and infant drowned together when she tried to save him. And when Washoe, the world's first language-trained chimp, heard a female scream and hit the water, she raced across two electric wires that normally controlled the apes to reach the victim, who was wildly thrashing about. Washoe waded into the slippery mud at the edge of the moat, grabbed one of her flailing arms, and pulled her to safety. She barely knew the victim, having met her only hours before.[45]

Obviously, intense hydrophobia cannot be overcome without an overwhelming motivation. Explanations in terms of mental calculations ("If I help her now, she will help me in the future") don't cut it: why would anyone risk life and limb for such a shaky prediction? Only immediate emotions can make one throw all caution to the winds, as empathy does by linking the emotional states of two individuals. In the words of the American psychologist Martin Hoffman, empathy has the unique property of "transforming another person's misfortune into one's own feeling of distress."[46] This mechanism was tested out, not on primates or other large mammals, but on rodents, by Inbal Ben-Ami Bartal at the University of Chicago. Bartal placed one rat in an enclosure, where it encountered a small transparent container, a bit like a jelly jar. Squeezed inside it was another rat, locked up, wriggling in distress. Not only did the free rat learn how to open a little door to liberate the

The empathy of laboratory rats has been tested by presenting them with a companion trapped in a glass container. Responding to the distress of the trapped rat, the free rat makes a purposeful effort to liberate her. This behavior disappears if the free rat is put on a relaxing drug, which dulls her sensitivity to the other's emotional state.

other, but she was remarkably eager to do so. Never trained on it, she did so spontaneously. Then Bartal challenged her motivation by giving her a choice between two containers, one with chocolate chips—a favorite food that they could easily smell—and another with a trapped companion. The free rat often rescued her companion first, suggesting that reducing her distress counted more than delicious food.[47]

Is it possible that these rats liberated their companions for companionship? While one rat is locked up, the other has no chance to play, mate, or groom. Do they just want to make contact? While the original study failed to address this question, a different study created a situation where rats could rescue each other without any chance of future interaction.[48] That they still did so confirmed that the driving force is not a desire to be social. Bartal believes it is emotional contagion: rats become distressed when noticing the other's distress, which spurs them into action. Conversely, when Bartal gave her rats an anxiety-reducing drug, turning them into happy hippies, they still knew how to open the little door to reach the chocolate chips, but in their tranquil state, they had no interest in the trapped rat. They couldn't care less, showing the sort of emotional blunting of people on Prozac or pain-killers. The rats became insensitive to the other's agony and ceased helping. This outcome fits far better with the idea of empathy-based helping, or sympathy, than with explanations based on immediate self-interest.[49]

The word *immediate* is critical here, because no one is claiming that empathy lacks any purpose in the long run. In biology, we distinguish sharply between two ways of serving one's own interests. First, at the evolutionary level, empathy would never have evolved if it didn't offer an advantage: it contributes to a cooperative society in which individuals can count on each other. Empathy probably brims with mutual benefits and survival value. The second meaning of self-interest is psychological, the goals that an individual pursues. Evolutionary goals are often unknown to individual actors. In the same way that young birds follow their species's migratory routes without knowing why, or that animals engage in sex unaware of its reproductive consequences, nature is full of evolved benefits that are not part of motivation. Psychologically speaking, animals can therefore be perfectly unselfish. If we scratch Washoe, the ape who rescued a drowning female, or Mae Perm, the elephant who guided her blind buddy, it is unlikely that we'll see bleeding hypocrites. Rather, we'll discover two kind souls, highly sensitive to the predicament of others.

Nevertheless, scholars keep obsessing about selfish motives, simply because both economics and behaviorism have indoctrinated them that incentives drive everything that animals or humans do. I don't believe a word of it, though, and a recent ingenious experiment on children drives home why. The German psychologist Felix Warneken investigated how young chimpanzees and children assist human adults. The experimenter was using a tool but dropped it in midjob: would they pick it up? The experimenter's hands were full: would they open a cupboard for him? Both species did so voluntarily and eagerly, showing that they understood the experimenter's problem. Once Warneken started to reward the children for their assistance, however, they became *less* helpful. The rewards, it seems, distracted them from sympathizing with the clumsy experimenter.[50] I am trying to figure how this would work in real life. Imagine that every time I offered a helping hand to a colleague or neighbor—keeping a door open or picking up their mail—they stuffed a few dollars in my shirt pocket. I'd be deeply offended, as if all I cared about was money! And it would surely not encourage me to do more for them. I might even start avoiding them as being too manipulative.

It is curious to think that human behavior is entirely driven by tangible rewards, given that most of the time rewards are nowhere in sight. What are the rewards for someone who takes care of a spouse with Alzheimer's? What payoffs does someone derive from sending money to a good cause? Internal rewards (feeling good) may very well come into play, but they work only via the amelioration of the other's situation.

They are nature's way of making sure that we are other-oriented rather than self-oriented. If we call this selfishness, then the word loses all meaning. When it comes to other species, too, the notion that all they seek is self-gain is an insult to their sociality.

Human beings evolved to reverberate with the emotional states of others, to the point that we internalize, mostly via our bodies, what is going on with them. This is social connectivity at its best, the glue of all animal and human societies, which guarantees supportive and comforting company.

4 | EMOTIONS THAT MAKE US HUMAN

Disgust, Shame, Guilt, and Other Discomforts

Queen Victoria may have been disgusted by the apes she met at the London Zoo, but what about the apes themselves? Are animals ever disgusted, and by what? When dogs lick their testicles, eat feces, or roll in stinky mud, we take this to mean that they lack any sense of shame or disgust. But the same argument could apply to our own habits. We like to eat oranges and squeeze fresh lemons, for example, but as soon as we offer a citrus fruit to our dog (not recommended), we see a full-blown disgust response, with curled-up lips, drooling, and withdrawal from the sour smell. A fruit that we view as healthy is revolting to another species. Do dogs ever wonder if humans know disgust?

Repulsed reactions are common in the apes. In our chimpanzee colony at Yerkes, the always-intrepid Katie was once digging into the dirt underneath a large tractor tire when she pulled out something wriggly. She uttered soft *hoo* alarm calls and held the object away from herself between her index and middle fingers, the way people hold a cigarette. First, she sniffed it, then turned and showed it to the others, including her mother, holding it high up with her outstretched arm like "Take a look at this!" It was probably a dead rat covered by maggots. Her mother gave a couple of loud *woaow* barks.

Recognizing the dramatic value of such objects, Katie's younger cousin, Tara, developed the naughty habit of carrying the corpse of a rat around by its tail, careful to keep it away from her own body, and surreptitiously placing it on the back or head of a sleeping group mate. Her victim would jump up as soon as she felt (or smelled) the dead animal, loudly screaming and wildly shaking her body to get this ugly thing off her. She might even rub the spot on her body with a fistful of grass, to make sure the smell was gone. Tara, for her part, would quickly pick up the rat and move on to her next target. Apart from the question of why Tara considered this game fun, and why we humans immediately see its humor value, my interest here is in the emotion of disgust, which has a mixed reputation.

On the one hand, disgust is considered evolutionarily primitive. Because it is often based on smell and serves to prevent the ingestion of harmful foods (citrus fruits are poisonous to canines), disgust is considered a basic emotion, sometimes even the "first" emotion. On the other hand, a burgeoning literature on disgust considers it a distinctively human sensation, one that is culturally constructed, used for moral dis-

The nose-wrinkle of disgust is a universal human expression, usually combined with narrowed eyes and knitted brows (*right*). It is a reaction to stinky food or other unpleasant situations but also expresses disapproval of human misbehavior. Chimpanzees have the same expression, such as their so-called rain face, which a relaxed female (*left*) pulls in response to a downpour (*middle*).

approval and the like. For example, the American neuroscientist Michael Gazzaniga in *Human: The Science Behind What Makes Us Unique* classifies disgust as one of the five emotional modules that distinguish us from all other animals.

Tara must not have read this book.

A Thirsty Horse

For anybody wondering which emotions make us human, I used to go with the most self-conscious ones, such as shame and guilt, even though I realize that some colleagues would go much further. They'd argue that animals possess only a handful of emotions, never mix them, and don't feel them the way we do. All this is pure speculation, though — of the same order as when José Ortega y Gasset out of the blue asserted that the chimpanzee differs from us in that he wakes up every morning as if no chimpanzee had ever existed before him. Was the Spanish philosopher suggesting that every chimp thinks he was created overnight? Why even say such a thing? In their desire to set humans apart, serious scholars come up with the craziest proposals — some fabricated, others unverifiable. We must take them with lots of salt, including those concerning what animals feel or don't feel.

Nevertheless, I was willing to sacrifice shame and guilt on the altar of the "only humans have X" religion that still reigns in academic circles. Both emotions require a level of self-awareness, I thought, that other species might lack. But I'm not so sure anymore. More and more I believe that all the emotions we are familiar with can be found one way or another in all mammals, and that the variation is only in the details, elaborations, applications, and intensity. Part of the problem is human language. You'd think it's an enormous advantage to be able to describe what you feel, but it's a mixed blessing that has landed the study of emotions into deep trouble.

It started with Ekman's labeling of facial expressions, presenting subjects with photographs of a face and asking them if it signals "anger," "sadness," or "joy." Seeing a picture of a laughing woman, you will without hesitation pick "joy" as the match. Done all over the world, people agree on a limited set of emotions. All this seems perfectly logical and informative. But what if you don't offer the subjects any labels to work with and just have them guess the emotion in their own words? What if you give them a list of labels that excludes the most obvious one—will they switch to an alternative? And how about photos of faces not taken under perfect light conditions? Actors make stereotyped faces, such as a laugh that cannot be mistaken for anything else. But "in the wild," humans make faces that are far less stereotyped, more fleeting, often of low intensity. We engage in subtle expressions while looking away, chewing food, blinking our eyes, sitting in the dark, and so on. Having done lots of additional research, it's no longer so clear how we interpret facial expressions. Given freedom to describe what they see, subjects do not always judge them in the standard way. There are a few on which the majority agrees, but the outcome is not nearly as homogeneous as once thought.[1]

Moreover, the labeling of emotions is a rather meaningless exercise, because emotions exist outside language. Chatting with a good friend over a cup of coffee on a sunny terrace, I will react within milliseconds to every facial or bodily move he makes without ever needing to search my mind for a word that goes with it. Humans constantly react to each other's body language in a stream, or "dance," of coordinated movements. While my friend talks, I raise my eyebrows, roll my eyes, mumble *hmmm* or *tsch*, and indicate with subtle muscle pulls around my eyes and mouth whether I agree, disagree, sympathize, approve, am amused, am surprised, and so on. My pupils dilate in synchrony with my friend's, and my body posture more often than not matches his. But if you asked me afterward what kind of faces my friend showed, I might not know or even care, because verbal labeling is not part of emotional communication. Language helps us discuss sentiments, but it doesn't play much of a role in how they are generated, expressed, or felt. Yet modern emotion research has placed language front and center.

Then there is the context of the facial expressions. Seeing a zoomed-in photo of tennis star Serena Williams, with wide open mouth and bared teeth, you'd think she was mad at her opponent. But the opponent happens to be her sister, Venus, whom she loves dearly and has just beaten in a match, which means she's probably ecstatic and may be screaming in triumph. This difference, which is rather crucial, is hard to tell from a close-up alone. Or you see a teary-eyed woman but can't tell if they're tears of joy at a wedding or tears of sorrow at a funeral. Is Uncle Joe baring his teeth in a photo in an attempt to smile or because he is strenuously uncorking a bottle of wine?

That faces are best judged in context has been taken to its extreme by Lisa Feldman Barrett, an American psychologist, who claims that emotions are mentally constructed. Instead of us being born with a set of well-defined emotions marked by clear body signatures, she argues, what we feel boils down to how we evaluate the situation we find ourselves in. Her position clashes with scientists who believe in Ekman's six basic emotions as the foundation of everything. The basic-emotions school loves simple labels for recognizable emotions, whereas Barrett is impressed by the variability of how we judge our feelings, which isn't always clear from how we express them.

People smile while they're sad, scream while they're happy, and even laugh while in pain. In a popular scene from the 1970s *Mary Tyler Moore Show*, Mary can't stop laughing at a funeral even though (or perhaps precisely because) she knows it is inappropriate. That the expressed exterior and the felt interior don't perfectly match doesn't mean that either is suspect, however. There is no great contradiction between assuming universal human facial expressions, understood all over the world, and acknowledging the absence of a one-on-one relation between expressions and feelings. The two don't always agree and don't need to.

For the same reason, I reject the notion that we're not allowed to speak of animal emotions since we can't know what they feel. One champion of the study of fear, who taught the world that fear runs via the amygdala, recently got so carried away by this argument that all of a sudden he refuses to speak of "fear" in the rats he has studied all his life. The American neuroscientist Joseph LeDoux has made ample comparisons between scared rats and human phobias, often using the words *rat* and *fear* in the same sentence. Now, however, he asks us to avoid any allusion to animal emotions, because emotional terminology cannot be employed without implying that rats feel the way we do. LeDoux, moreover, surmises that since we have dozens of different words related to fear (*phobia, anxiety, panic, worry, dread,* and so on), and since rats don't have all those words or any at all, it's impossible for them to experience as many shades of emotion as we do.[2]

This argument, which posits that language lies at the root of emotions, reminds me of an encounter during a workshop on sexual behavior, in which postmodern anthropologists put more trust in language than in the scientific method. Feelings can't be felt without words, they argued, going so far as to claim that it is impossible for people whose language lacks a word for *orgasm* to feel any sexual pleasure. This unsupported claim so upset the scientists present that we began circulating little notes to each other with questions such as "If people lack a word for oxygen, can they breathe?" Emotions clearly precede language in both evolution and human development, so language can't be all that important. It's tacked on. All it does is label internal states, but who says it helps us distinguish them? The German language has two different words for anger and disgust, whereas Yucatec Maya, a language from Mexico, covers both emotions with a single word, yet people from both cultures are equally good at distinguishing faces expressing anger or disgust. Knowledge of the emotions clearly goes beyond words.[3]

But LeDoux has become so afraid of the term *fear* that he now denies this emotion in his rats. Instead, he grants them "survival circuits" in the brain that make them respond to existential threats. I'm extensively familiar with this argument, because ethology (the European school of animal behavior in which I was trained) preferred similar functional interpretations. It didn't want to touch internal processes with a ten-foot pole. My ethology professors would literally pull disgusted faces—an emotional expression shared with other animals!—as soon as the word *emotion* came up in relation to animals. They felt much more comfortable with functional stories about how a given behavior assists survival.

To return to rat emotions, we've known all along that emotions and feelings are different things. Emotions are bodily expressed, hence observable, whereas feelings remain private. Nothing new here. So why do we hear only now that since we can't know what rats feel, it's best to avoid any talk of their emotions? And why not then extend the same argument to our own behavior? We may have lots of words for fear, but does this truly help us understand this state in one another? Do we know the exact meaning of all these words, and do they adequately cover what is being felt? Is our vocabulary up to the task? If I ask you how you felt about your father's death, for example, you may tell me that you were "sad," but does this really allow me to grasp your feelings? I can't get inside you. Who says your sadness is like my sadness, and who says yours wasn't mixed with relief, or perhaps anger, or some other feeling you'd rather not talk about? There may even have been emotions involved that you won't admit to yourself.

Emotions often stay subconscious. As a student, I was about to make my very first airplane flight, to see orangutans in the rain forest of Sumatra, Indonesia. You might think I was worried about snakes and tigers, or perhaps the millions of leeches that crawl over the forest floor, but I was very much looking forward to my first tropical trip. Or at least, I thought I was. The closer the date came, however, the more I had belly problems. I didn't know where they came from, but my stomach was in a knot for weeks up to and including the day I stepped onto the plane. My symptoms miraculously vanished, however, once the plane landed in Medan. A day later I arrived in the jungle in the best mood and had a great stay. In retrospect, I concluded that I had been scared to death of the plane ride but had suppressed this feeling since it would have interfered with my dream to see wild orangutans. I don't think I'm the only one who strains his prefrontal cortex to block awareness of inconvenient emotions. What humans tell us about their feelings is often incomplete, sometimes plainly wrong, and always modified for public consumption.

As if this weren't enough of a problem, even the best, most accurate descriptions still can't make me feel what you feel. Feelings are private experiences, and we can talk about them all we want, but they remain private. I doubt therefore that I know the feelings of my fellow human beings any better than I know the feelings of the animals I work with. It may seem easier for me to extrapolate from your feelings than from a chimp's, but how would I know for sure? Unless, of course, we assume that animals have no feelings at all. In this case, we could go with LeDoux's proposal and skip the implication of felt emotions altogether. But this is a highly unreasonable position given how similarly the emotions manifest themselves in animal and human bodies and how alike all mammalian brains are down to the details of neurotransmitters, neural organization, blood supply, and so on. It would be like saying that both horses and humans seem to get thirsty on a hot day, but in horses we should call it "water need" because it is unclear that they feel anything. This then raises the question how a horse decides it needs a drink, if not for signs of dehydration inside its own body. The horse's body detects internal changes and sends information to its hypothalamus, which monitors the sodium concentration in its blood. If this level rises above a certain limit, the blood gets too salty, and its brain induces a strong desire for gulps of water. Desires work by being felt. The horse will be irresistibly drawn to a river or water trough. This detection system is one of the oldest in existence and is essentially the same across many species, including our own. Does anyone truly believe that after a long trip through the desert, the cowboy feels differently about water than his horse?

I'm all for calling horses thirsty and rats fearful based on their behavior and the circumstances under which it arises, while fully realizing that I can't feel what they feel—I can only guess. This situation is not radically different, to my mind, from that concerning human emotions. When it comes to feelings, all I know for sure is what I feel myself, and even there I mistrust my impressions given how prone I am to wishful thinking, denial, selective memory, cognitive dissonance, and other mental trickery. Most of us are not like the French novelist Marcel Proust, who continually analyzed his own sentiments and became intimately familiar with them. But even Proust concluded (about a romantic partner whom his protagonist no longer loved, until she died and he realized he still did) that "I had been mistaken in thinking that I could see clearly into my own heart."[4] He couldn't, because the heart often knows better than the mind what we feel. I realize this is a rather unscientific take on the heart, and it may be better to refer to the body as a whole, but it is undeniable that we have trouble penetrating our own emotional life. This doesn't keep us from dissecting and discussing it all the time, however, spending tons of imprecise words on a most slippery topic, which makes the timidity of science about animal emotions seem all the more out of proportion.

An Eye for an Eye

Science often compares adult apes to children, as in "the chimpanzee has the mind of a four-year-old." I never know what to do with such statements, though, given that I find it impossible to look at an adult chimpanzee as a child. A male is interested in power and sex and prepared to kill for it. If he is high-ranking, he may adopt a leadership role, which includes keeping order and defending the underdog. Males engaged in power struggles sometimes have a permanently furrowed-brow expression suggesting internal turmoil, and they are known to have high levels of stress. A female ape, on the other hand, is mainly interested in her offspring and the duties that come with motherhood, such as taking time to nurse, finding food, and deterring predators and aggressive members of her own species. She also works every day on her relationships, grooming her friends, consoling them after upheavals, and watching over their offspring if needed. The lives of adult apes center very much on adult concerns, therefore, and share little with a child's insouciance.

Juvenile apes squabble over food and hit each other screaming over the head, while adult apes politely beg and share, sometimes taking turns, while exchanging food for services received earlier in the day. Here, too, the best comparison is between ape and human juveniles or ape and human adults. This matters in relation to emotions, because some emotions are typical of adults, especially those that require a greater appreciation of time than is found in the young. Youngsters live in the moment, while adults don't. Some emotions are future oriented, such as hope and worry, while others relate to the past, such as revenge, forgiveness, and gratitude. All these *time-line emotions,* as I like to call them, seem present in adult apes and in some other animals as well.

In chimpanzees, sharing food is part of a give-and-take economy that includes grooming, sex, support in fights, and other types of aid. All these favors are thrown into one big exchange basket glued together by the emotion of gratitude. Gratitude functions to maintain balance sheets of exchange: it prompts individuals to seek out those who have been good to them, and—if the occasion arises—return their favors. Based on thousands of observations, we have found that chimpanzees share food specifically with those who have been kind to them in the past. Every morning, when the apes gather in the climbing frame to patiently tend each other's hair, we measure who grooms whom. In the afternoon, we provide them with shareable food, such as a few large watermelons. Melon owners allow anyone who has groomed them to remove pieces from their hands or mouth, but not individuals with whom they failed to interact in the morning—they may resist the latter individuals and sometimes even threaten them. Sharing patterns thus change from day to day depending on the distribution of earlier grooming. Since the time span between the two events is several hours, the sharing requires memory of past encounters and positive feelings about enjoyed services. We know this combination as gratitude.[5]

Mark Twain once quipped, "If you pick up a starving dog and make him prosperous, he will not bite you. This is the principal difference between a dog and a man." In my own home, adopted stray pets have always been most grateful for the warmth and food we offer. A scrawny kitten full of fleas, picked up in San Diego, grew into a gorgeous tomcat by the name of Diego. Diego purred excessively his entire fifteen-year-long life whenever he was fed—even when he barely ate anything. We interpreted his behavior as gratefulness, but it is hard to exclude mere happiness. Diego may just have enjoyed food more than your average spoiled pet.

In apes, signs of gratitude may be more obvious. Two chimps had been shut out of their shelter during a rainstorm. Wolfgang Köhler, the German pioneer of tool use studies, happened to walk by and found both apes soaking wet, shivering in the rain. He opened the door for them. Instead of hurrying past him to enter the dry area, however, the chimps hugged the professor in a frenzy of satisfaction.[6]

Their reaction resembles that of Wounda, a chimpanzee who had been rescued from poachers, near death, and received medical care at the Tchimpounga Rehabilitation Center in Congo-Brazzaville.[7] In 2013 she was released back into the forest. A video of this moment went viral because of the emotional interaction between Wounda and Jane Goodall, who attended her release. At first Wounda walked away, but she then hastily returned to hug the people who had taken care of her. She specifically turned to Goodall for a long mutual embrace before taking off. This was remarkable because Wounda had first been on her way, then seemed to correct herself and return, as if realizing it wouldn't be very nice to just walk away from those who had saved her and nursed her back to health.

Similar accounts exist of netted or beached dolphins and whales that human divers cut loose from nets or pushed back into the ocean. The cetaceans returned to their rescuers and nudged them or lifted them half out of the water before swimming away. In all cases, the humans present, deeply moved, viewed these interactions as signs of gratitude.

I have already mentioned how Kuif, Mama's best friend, was affected by my teaching her how to raise a baby on a bottle. From the moment we gave her permission to pick up Roosje, the adoptive baby we put in the straw of her bedroom, she treated me as family, something she'd never done before. I saw this as a sign of gratitude for changing her life for the better, from a mother who had lost several infants due to lactation failure to one who successfully raised Roosje and later applied the same bottle skills to her own offspring.

The ugly sister of gratitude is revenge, an emotion equally concerned with the squaring of accounts, but in a negative sense. Edvard Westermarck, the Finnish anthropologist who gave us the first ideas about the evolution of human morality, stressed the value of retribution in keeping people in line. He didn't think we were the only species with this tendency, but in his time, there was little animal behavior research. So he relied on anecdotes, such as one he heard in Morocco about a camel that had been excessively beaten by a fourteen-year-old boy for turning the wrong way. The animal passively took his punishment, but a few days later, while unladen and alone on the road with the same conductor, it "seized the unlucky boy's head in its monstrous mouth, and lifting him up in the air flung him down again on the earth with the upper part of the skull completely torn off, and his brains scattered on the ground."[8] Stories of resentful animals can often be heard at zoos, usually concerning elephants (with their proverbial memories) and apes. Every new student or caretaker working with apes needs to be told that they won't be able to get away pestering or insulting them. An insulted ape remembers and will take all the time in the world for the right opportunity to get even. Sometimes it doesn't take long. One day a woman came to the front desk of a zoo where I worked complaining that her son had been hit by a rock coming from the chimpanzees. The son was surprisingly subdued, however. Witnesses later told us he'd thrown the same rock first.

Among themselves as well, chimps retaliate. They support each other in fights, following the rule that one good turn deserves another; experiments have confirmed this tendency. Given a chance, many animals are prepared to do a partner a favor, such as by pulling a lever or selecting a token that produces food. They do so at a moderate level as long as their partner is a passive recipient, but they dramatically increase their generosity when the partner is allowed to return the favor. When both parties stand to gain, they move to a new stage. This is of course also how things work in real life.[9] What may be unique for chimpanzees is that they apply similar tit-for-tat rules to negative acts. They tend to get even with those who have acted against them. For example, if a dominant female often attacks another female, the latter cannot retaliate on her own but will wait for the best occasion. As soon as her tormenter is involved in a brawl with others, she will join in to add to her troubles.

After Nikkie became the new alpha male in the chimpanzee colony of Burgers Zoo, he regularly practiced strategic retaliation. His dominance was not yet fully acknowledged, and subordinates would often pressure him, banding together and chasing him around, leaving him panting and licking his wounds. But Nikkie did not give up, and a few hours later he'd regain his composure. The rest of the day he'd go around the large island to single out members of the resistance, visiting them one by one while they were sitting alone minding their own business. He'd intimidate them or give them a beating, which likely made them think twice before opposing him again. This eye-for-an-eye tendency is so prominent in chimpanzees that it can be statistically demonstrated in thousands of observations in our database.

Retaliation is an "educational" reaction that attaches costs to undesirable behavior, but it is unclear if the apes themselves think along these lines.[10] They probably just follow an urge to take revenge, a tendency we share with them. After all, we call revenge "sweet," as if it were something delicious. When experimenters gave human subjects voodoo dolls representing people who had insulted them, their mood improved markedly when they were allowed to stick needles into those dolls.[11] Our judicial systems take the longing to get even a step further: when the family of a murder victim or those who have been scammed out of money seek redress, they are undeniably driven by a deep desire to inflict harm on those who have done harm to them.

Chimpanzees do the same, thanks to their flexible hierarchy that offers room for retaliation. Rhesus macaques and baboons, in contrast, have such despotic hierarchies that it is almost suicidal for a subordinate to turn against a superior. Intimidation and punishments always flow down the hierarchy, which excludes opportunities for revenge. But even these monkeys know how to hit back: they rely on the kinship bonds that pervade their society. Grandmothers, mothers, and sisters spend extraordinary amounts of time together, forming tight units known as matrilines. A monkey who has become a victim of aggression will likely vent his feelings on a relative of the attacker. Instead of retaliating, which he can't, he will look for a younger member of the attacker's matriline, who will be easier to intimidate. They sometimes perform vindictive actions after quite a delay, suggesting they have excellent memories.[12] This tactic obviously requires monkeys to be aware of the family to which every other monkey belongs, and we know they are. It would be as if I responded to a reprimand from my boss by yanking the hair of his young daughter. I don't go against the pecking order, but I still punish my offender.

The final emotion concerning past events is forgiveness. Having studied primate reconciliation all my life, I have seen many times how chimpanzees kiss and embrace their former adversaries, how monkeys groom them, and how bonobos resolve social tensions with a little sex. This kind of behavior is not at all limited to primates: hundreds of reports find it in other social mammals and in birds, so much so that if anyone were to claim that a given species *doesn't* make up after fights, we'd be baffled.

Conflict resolution is part and parcel of social life. The emotions involved are hard to pinpoint, but a minimum requirement is that anger and fear—the typical emotions during a confrontation—are toned down in order to permit a more positive attitude. This reversal is rather counterintuitive. Someone who has just lost a fight with a dominant attacker now needs to pluck up the courage to approach him or her for a friendly reunion. Meanwhile the aggressor must suddenly drop the enmity, which is illogical. But many animals undergo these emotional changes remarkably rapidly, as if a control knob in their mind were turned from hostile to friendly.

Humans become masters at turning the same emotional knob if we live in a conflict-prone environment, such as a large family or a workplace with lots of colleagues. These places require compromise and forgiveness every day. Forgiveness is never perfect, though, and even though we often say "Forgive and forget," the forgetting part is problematic. We don't erase the memory of a slight but simply decide to move on. Many group-living animals do the same because they, too, depend on peaceful coexistence and cooperation. Reconciliation is the observable process, whereas forgiveness is how it is internally experienced. Given the evolutionary antiquity of this mechanism, it is hard to imagine that the emotions involved are radically different between us and other species.

All three emotions—gratitude, revenge, and forgiveness—sustain social relationships based on years of interaction among individuals, sometimes going back to when they played together as youngsters. These emotions serve friendships and rivalries, enhance or harm trust, and keep the society functioning to everyone's advantage. Animals are remarkably good at this balancing act, which requires them to exchange favors and resolve tensions. We know now that monkeys (and probably other animals as well) have dedicated brain networks that help them process social information. These neural networks have been tested: while monkeys watch televised scenes and see their fellows engage in social affairs, they are activated, but they remain inactive during physical or ecological scenes. Students of animal behavior have for a long time insisted on the special status of social intelligence: neuroscience is now backing us up.[13]

Are there also emotions concerned with the future? It is well established that apes and some large-brained birds do not live purely in the present. Wild chimpanzees plan ahead by picking up tools hours before they arrive at the termite hill or beehive where they will use these implements. While collecting them, they must know where they are going. Similar planning has been demonstrated in primates and corvids, showing they may ignore immediate gratification to reap future benefits.[14] Given a choice, apes will forgo a juicy grape placed right next to a tool that they can use only hours later for a better reward. This takes self-control. Planning is harder to prove in the social domain, even if the political battles among male chimpanzees are suggestive. When a young adult male challenges the established boss, he may lose about every confrontation and sustain frequent injuries. Yet he will keep going day after day without any immediate rewards. Only months later he may finally have a breakthrough and get support from others who help him topple his adversary. And even then, as was the case with Nikkie, the young male may still meet resistance before he is fully accepted. It may take years before his position is truly secure. Was this his plan all along? And if not, why go through this hell? It is hard to watch these strategies, as I have done so many times in my career, and not think that they are built on hope.

"Hope" is rarely attributed to animals, but the related idea of "expectation" was proposed already a century ago. The American psychologist Otto Tinklepaugh conducted an experiment in which a macaque watched a banana being placed under a cup. As soon as the monkey was given access to the room, she ran to the baited cup. If she found the banana, everything proceeded smoothly. But if the experimenter had surreptitiously replaced the banana with a piece of lettuce, the monkey could only stare at it. She'd frantically look around, inspecting the location over and over, while angrily shrieking at the sneaky experimenter. Only after a long pause would she settle for the disappointing vegetable. Tinklepaugh demonstrated that instead of a simple association between location and reward, the monkey recalled what she had seen being hidden. She had an expectation, the violation of which greatly upset her.[15]

Primates and dogs react with similar surprise when human magicians make things miraculously disappear or conjure them out of thin air. Apes may laugh or look puzzled, while dogs madly search for the vanished treat, indicating that the reality they had in mind was different.

Expectations also feed tit-for-tat barter, which is well known among animals despite Adam Smith's claim that "nobody ever saw a dog make a fair and deliberate exchange of one bone for another with another dog."[16]

Bartering is second nature to primates. Here an adolescent bonobo has noticed that an adult male holds two grapefruits in his hands. She hurries over to present for sex, showing an orgasm face during copulation. Afterward the male shares one of his fruits with her.

Smith may have been right about dogs, but wild chimpanzees in Guinea are known to raid papaya plantations in order to buy sex. Adult males usually steal large fruits, one for themselves, the other for a female with a genital swelling. The female waits at a quiet spot while the male risks the anger of farmers in order to bring her a delicious fruit, which he hands to her during or right after sex.[17]

In another instance of tit for tat, the long-tailed macaques at some Balinese temples have a habit of stealing valuable objects from tourists. At temple entrances, written signs warn everyone to remove their glasses and take off their jewelry, but many tourists fail to do so, not realizing how incredibly fast the monkey mafia operates in these places. A monkey will jump onto their shoulder to snatch a pair of glasses or run off with a precious smartphone. They will steal flip-flops by literally removing them from the tourists' feet. Instead of playing with these items or taking off with them, the monkeys sit down patiently nearby, but out of reach, to see how much their victim is willing to pay to get an item back. A few peanuts won't do. The monkeys want at least a whole bag of crackers before they will drop the item. Primatologists who studied this extortion game found that the monkeys had a pretty good idea of which objects humans value the most.[18]

Given the existence of future-oriented behavior, dogs have recently been classified as either "optimistic" or "pessimistic" while facing a given task. Dogs that get seriously upset when their owner leaves them alone at home—venting their frustration by destroying the house, relieving themselves, or furiously barking—are regarded as pessimistic. When presented with a food bowl of unknown content, they hesitate and approach slowly, perhaps expecting the bowl to be empty. Dogs that are less perturbed by separation, on the other hand, are considered more optimistic: they happily run toward the bowl, expecting it to be full. This so-called *cognitive bias* is also common in people. Cheery, easygoing people expect good things in life, whereas depressed ones believe that everything that can go wrong will go wrong.[19]

Cognitive bias offers us a rare opportunity to test how farm animals feel about their living arrangements. After all, pigs that live under stress in small crates may expect few good things to happen to them. But those that live in a fun environment, with straw to burrow into while sleeping in piles, enjoying body heat and physical contact, may be in better spirits. In one study, groups of pigs were housed either in small pens with concrete floors or in larger pens with fresh straw every day and cardboard boxes to play with. All the pigs were trained on two different sounds: a positive sound would announce a slice of apple, and a negative sound announced only a plastic grocery bag waved in the pig's face. Being intelligent, the pigs quickly learned to go for the positive sound.

After this training, the pigs were presented with an ambiguous sound, somewhere between the other two. What would they do? It depended entirely on their living conditions. The pigs in the enriched environment expected good things to happen and eagerly approached the ambiguous sound, but the ones kept in the barren environment did not see things the same way. They stayed away, perhaps expecting that stupid plastic bag again. If their housing was changed for better or worse, the pigs' responses to the ambiguous sound followed suit, indicating that their daily life affected how they perceived the world. The cognitive bias test is informative, allowing us to verify the claims of companies that advertise their products as coming from happy animals, such as the well-known French spreadable cheese *La vache qui rit* (the laughing cow). The test can tell us if these animals really have anything to laugh about.[20]

The condition of looking forward to something desirable is what we call "hope." A monkey looking for a lucrative trade, a chimpanzee trying to improve his status, a dolphin searching the ocean for her lost calf, wolves setting out on a hunt, or a herd of elephants following an old matriarch who knows the last watering hole in the desert—all may well experience hope. It may be present or absent in farm animals as well. Like us, many animals evaluate everything that happens to them against a backdrop of past and future. Timeline emotions that transgress the present can no longer be denied, given the mounting evidence that animals hold memories of specific events, are forward-looking, exchange favors, and engage in an eye-for-an-eye.

Pride and Prejudice

The Jamaican sprinter Usain Bolt celebrated his wins with a "lightning bolt" display, bending one arm at the elbow while stretching his other to point in the distance. Many prominent people have imitated his signature victory stance. Famous European footballers, after scoring a goal, lift their shirts to show off their abs while sliding on their knees along the grass, with arms spread, to absorb the adulation of the roaring crowd.

Athletes celebrate victory by expanding their bodies and raising their arms to communicate that they are the winner. This expression of pride is a human universal. Similar victorious signals are known of animals who have defeated a rival, such as the so-called triumph display of geese.

Generally, we make ourselves look bigger after a win, we expand, in what amounts to a triumph display: chin lifted, chest out, shoulders pulled back, arms moved away from the body, and always that smile. The corresponding emotion is pride. In animals it is usually called dominance, but the principle is the same: animal winners make themselves look larger by putting up their feathers or hair, walking with legs apart, holding their heads high, stretching their torsos, and so on. This puffery creates an illusion of greater size, so that one might erroneously think it's always the largest who wins.

Caitlin O'Connell, an American expert on elephants, writes of Greg, the top-ranking bull elephant around Etosha National Park in Namibia:

There is something deeper that differentiates him, something that exhibits his character and makes him visible from a long way off. This guy has the confidence of royalty—the way he holds his head, his casual swagger: he is made of kingly stuff. And it is clear that the others acknowledge his royal rank as his position is reinforced every time he struts up to the water hole to drink.[21]

Power signals by dominant individuals go back far in evolutionary time. Fish make threat displays by spreading all their fins. Some lizards expand the frills around their neck, and the dominant rooster in a flock crows first. Perhaps best known of all is the *Triumfgeschrei* (German for "triumph-screaming") of greylag geese. The gander, having chased off an intruder, runs back to his mate with spread wings while uttering blaring calls. The two of them then engage in a bonding ritual to celebrate the defeat of their rival, in which they both stretch their necks in parallel and noisily call together. Their bond has survived yet another challenge.

The American psychologist Jessica Tracy documented the human triumph displays in her 2016 book *Take Pride: Why the Deadliest Sin Holds the Secret to Human Success.* She analyzed hundreds of photographs of

world-class athletes to determine how they reacted to winning or losing. Photographs of winners of the 2004 Olympic judo competition,
taken right after every match, showed the same pride expression: body
expansion, raised arms, and fisted hands. We often think Westerners are
more into individual success, stressing their own qualities and accomplishments, but in fact the cultural background of the athletes didn't
matter. All winners showed the same behavior regardless of nationality.
Again, we might ask if this is because the world has been homogenized.
Do athletes learn how to express triumph from watching each other?
Tracy addressed this question by analyzing a new set of photographs
taken during the Paralympic Games, showing congenitally blind athletes. Winning blind athletes celebrated in exactly the same manner as
seeing ones, so Tracy concluded that pride expressions are not acquired
from others. They are biological, an idea that is further supported by
their similarity with other species.[22]

But Tracy did not recognize the same pride emotion in animals;
instead she offered the sort of functional account that my ethology professors would have loved. The only reason animals make themselves
look large, she suggests, is to bluff, threaten, or intimidate. It's all a
means to an end. They show this kind of behavior before or during
confrontations, whereas humans do it *after* the defeat of a rival, hence for
a different reason. Only humans feel a sense of accomplishment, Tracy
concluded, which "requires an understanding that the self is a stable
entity that has continuity over time, that who I am and what I do now is
relevant to who I was yesterday and who I will be tomorrow."[23]

Do I hear an echo of Ortega y Gasset's claim that a chimp wakes up
every morning with no idea who or what he is? I'm baffled, because for
most animals, today's behavior is a direct continuation of yesterday's and
predicts tomorrow's. Imagine having to start over every day figuring out
their hierarchy and social network! In reality, each individual member of
an animal society has an enduring role. Like us, animals know exactly
who they are and where they fit in. Their friendships often last a life-

time. In addition, status displays are much more than a way to get the upper hand. Sometimes they relate to recent history, such as the triumph ceremony of geese, which typically follows a victory, such as a gander chasing off a rival. Doesn't this ceremony suggest the gander is proud, as it does with humans?

Similarly, a coyote who has successfully pinned another to the ground may emerge with a bouncing gait while the loser remains flat on his belly. After fights between domestic cats, the victor often demonstratively rolls around on his back within sight of the loser. In mangrove crabs, triumph displays are common after contests. The victorious male rubs one claw vigorously against the other, thus stridulating a song to celebrate his success.[24] In other animals, too, from wolves to horses to monkeys, you need only one glance at an individual to know if he or she has a history of winning or losing. It's written all over them. That Greg, the elephant, walks his majestic walk oozing self-confidence is obviously based on a track record of things going his way.

A chimpanzee alpha male has his hair almost permanently on end, making it easy to tell him apart from other males. He may show a "bipedal swagger," in which he walks upright on both legs with arms hanging loose, away from his body, his torso swaying from side to side as if he is top-heavy. He may also menacingly hold a rock or stick in his hand. It is such an arrogant pose, and so recognizable, that I often show appreciative audiences a picture of a swaggering upright male chimp next to the almost identical gunslinger walk by a former U.S. president from Texas.

I prefer the way Abraham Maslow thought about the attitudinal difference between high- and low-status primates. It is a little-known fact that long before this American psychologist became famous for his hierarchy of needs—now a staple of psychology textbooks and management training—he conducted graduate research on primate social dominance. I am very familiar with it as he worked at Vilas Park Zoo, in Madison, Wisconsin, where I observed macaques a couple of decades later. Maslow described the cocky, self-confident air of dominant monkeys, and the slinking cowardice, as he called it, of subordinates. A rhesus monkey alpha male walks the entire day with his tail in the air, and he is the only one to do so, although other males dare lift their tails when an alpha is out of sight. An alpha regularly bounces up a tree, vigorously shaking its branches, letting everyone know who's boss. Maslow's concept of "self-esteem" grew straight out of what in primates he termed "dominance feeling." Initially, he used both terms interchangeably, stressing how human psychology is rooted in monkey behavior. Maslow thus appreciated the self-assurance of high-status primates, including their sense of superiority.[25]

The difference between Tracy's and Maslow's views boils down to the level of self-consciousness we are willing to grant to other animals. I hate to put it this way, though, because if there is one capacity that escapes proper definition, it is consciousness. This means we have to work with assumptions, although not entirely, as observable behavior remains the starting point.

And in terms of observable behavior, the similarities in attitudes between humans and other animals are striking. Darwin noticed that an assertive dog's body posture (lifted head, stiff legs, bristling hair) is the opposite or "antithesis" of a submissive one's posture (crouching, tail down, hair sleek). Given that such status signals are universal, shouldn't we also assume the underlying emotions to be the same? Evolutionarily speaking, this is the safest bet whenever related species act alike. We don't want to burden pride with prejudice by postulating major emotional differences for which there is no evidence, which is why I side with Maslow's view that individuals who systematically outcompete others, whether they be humans or other animals, probably arrive at a radically different self-assessment. They feel good about themselves and flaunt it in their behavior. Not only do expressions of pride have a long evolutionary heritage, therefore, so do the associated emotions.

Guilty as a Dog

In Tracy's study, the losing judokas shrank in size, slumped their shoulders, and lowered their heads, exhibiting all the signs of shame and failure. This is also the typical reaction when people fail to meet expectations or anticipate trouble after having violated a norm. The word *shame* is thought to come from an earlier word that means "to cover." We lower our face, avoid the gaze of others, bend our knees, down our eyelids, and generally look miserable and diminished in stature. Our mouths droop, and our eyebrows arch outward in a distinctly unthreatening expression. We may bite or bulge our lips, or hide our face behind our hands as if we wanted "to sink into the ground." We say we're ashamed, but we also know that people are angry with us, or at least irritated and disappointed.

The diminished stature and desire for invisibility of an ashamed athlete have obvious parallels in primate submissive behavior. Chimpanzees crawl in the dust for their leader, lower their body so as to look up at him, or turn their rump toward him, which makes them vulnerable.

When pet owners come home and discover rule-breaking, such as a shredded pillow or chewed shoe, they have no trouble telling who did it. While the culprit dog (*right*) is being scolded, he refuses to look up and adopts a submissive posture. But even though he acts guilty, it is unclear if he actually feels remorse. More likely, he knows he is in trouble.

Dominant chimps may emphasize the contrast by literally walking over a subordinate, or running past him while moving a lifted arm over his back, giving them no other choice than to duck into a fetal position.

Notice the different language applied to humans and animals, though. We've already seen this with *pride*, the word used for humans, versus *dominance* in other species. Similarly, a person who gets into trouble with others or loses a contest is said to be *ashamed*, whereas a chimp under the same circumstances is merely *submissive* or acting like a *subordinate*. We prefer functional terms for animals, whereas for ourselves we focus on the feelings behind behavior. We are reluctant to imply that animals may have the same feelings or any feelings at all. But obviously emotions must be involved, and why would they be different? If shame were truly a uniquely human emotion without evolutionary antecedents, then shouldn't humans express it quite differently from animals? Why must it look exactly like the behavior that biologists would classify as submissive? And not only biologists: Daniel Fessler, an American anthropologist who specializes in human shame, compares its universal shrinking appearance to that of a subordinate facing an angry dominant. Shame reflects awareness that one has either upset others or made a fool of oneself, so that mollification and explanation are in order. Its hierarchical template is obvious.[26]

This doesn't mean human shame is exactly like submission. It certainly seems to have a wider reach in us than in other primates. I've never seen young chimpanzees ashamed of their mother, or chubby elephants obsessed with their weight. We humans are great at cultural habits, norms, and fashions, which change all the time, creating uniquely human grounds for shame, including intergenerational ones. Teenagers, for example, are embarrassed when their parents are out of tune with fashion or employ a vocabulary from ages (two decades) ago. The same teenagers have no problem with their parents at home, but as soon as their friends are around, things change: *What will they think if they see me walking with these Neanderthals?* At first sight, being ashamed of one's parents is about conformity rather than hierarchy, but in the end, it's all about the teen's reputation and standing within the peer group.

Only one shame display is new to our species and so hints at either a deeper or new emotion: blushing, a color change in the face and neck region that is produced by increased blood flow through subcutaneous capillaries. I've already mentioned that it is uniquely human. Charles Darwin was so puzzled by it that he wrote letters to colonial administrators and missionaries all over the world to see if humans everywhere blushed. He speculated about the effect of skin color (a reddening face stands out more against a lighter background), and the role of shame and moral standing. His main conclusion was that blushing was an innate, universal reaction in our species that evolved to broadcast shame or embarrassment.

Blushing is highly communicative yet involuntary. Even tears can be faked more easily than a blush. We are unable to produce it on command, and unable to suppress it if we wish it to go away. In fact, the more aware we are that we are blushing, the harder it is to make it disappear. Why does our species need a shame signal that other primates lack, and why did nature not grant us more control over it?

The main issue here is trust. We trust people whose emotions we can read in their faces over those who never show the slightest hint of shame or guilt. We have another characteristic that fits the same pattern: the white sclera around the eyes. They make our eye movements stand out more than those of, say, a chimpanzee, whose eyes are all dark and recessed in the shade of a prominent eyebrow ridge. There is no way to tell where a chimp is looking from the eyes alone, whereas humans have trouble obscuring their gaze direction or hiding the restless glancing that betrays nervousness. It hampers our manipulation of others. During human evolution, trustworthiness must have become such a premium that deceptive capacities had to be handicapped. It made us more attractive as partners. Blushing may be part of the same evolutionary package that gave us high-level cooperation and morality.

Shame also surrounds sex, as in our desire for privacy and the obligation to cover certain body parts in public. Some of this is entirely cultural. I for one have never got used to the American fixation on breasts. My first culture shock in this country was when I read in a morning paper that a woman had been arrested for breastfeeding in public. In the Netherlands, this would never be an issue, and moreover I am a primatologist, for whom there is nothing more natural than a baby on the nipple. But all over the world, humans have declared certain areas related to sex and reproduction to be off limits visually. At its most extreme, people are unable to make love with the lights on.

Some of these taboos are hard to fathom, but it probably all started with a need to protect the family. Human societies are characterized by units involving fathers and mothers, both of whom have a vested interest in safeguarding their bond. Instead of having a territory and keeping everyone else out, which is how birds and many other animals handle this issue, we live together with lots of potential sex partners and rivals. There is no shortage of extramarital affairs, of course, but also a need to keep them under control or at least under the radar. This is a major difference between us and other hominids, which have no nuclear families. Ape females raise offspring on their own. Even if some males and females prefer each other's company, their bonds are nonexclusive. For chimps, the only time sex has to be removed from public view is when a male and female worry about the jealousy of rivals. They will meet behind the bushes or travel away from the rest of the community, in a pattern that may be at the root of our desire for privacy. "Concealed copulation," as biologists call it, is a fairly common phenomenon in animals. Sex being a major source of competition and violence, one way to keep the peace is by restricting its visibility. Humans take this need a step further than chimps, hiding not only the procreative act but also covering up any arousing or arousable body parts, at least in public.

None of this occurs in bonobos, which is why these apes are often considered sexually "liberated." But since privacy and repression are nonissues in their highly tolerant societies, liberation is a nonissue, too. They simply have no modesty, no inhibitions other than the wish to avoid trouble with rivals. When two bonobos couple, the young sometimes jump on top of them to take a peek at the details. Or an adult may move in and press her genitals against one of the sex partners to take part in the fun. In this species, sexuality is more often shared than contested. A female may lie on her back masturbating in full view of everyone, and no one will blink an eye. She moves her fingers rapidly up and down her vulva, but she may also assign a foot to the job, keeping her hands free to groom her infant or consume a fruit. Bonobos are great multitaskers.

Emotionally close to shame we find guilt. But guilt is concerned with an action, whereas shame is more about the actor. *I shouldn't have done this!* is what a guilty person feels, whereas shame is more like *Don't look at me, I'm worthless!* Shame is concerned with judgment by the group, but guilt with judgment by oneself. Going by its external signs, however, the two emotions are hard to tell apart, and the animal parallels are equally strong for both. This is why many dog owners are convinced that their pets feel guilt. The Internet features multiple videos about two dogs, one of whom has eaten the kitty kibble and the other who is innocent. My favorite is "Denver, the Guilty Dog," in which Denver shows all the signs of realizing that punishment hangs over his head.[27] No one doubts that dogs know when they are in trouble, but whether they actually feel guilty is a point of debate.

One American expert, Alexandra Horowitz, tested things out by having dogs meet an angry owner when they had done nothing wrong or a relaxed owner when in fact they had messed up the kitchen or ruined some fancy shoes (or in the case of Horowitz's experiment, eaten a biscuit that the owner had told them not to touch). After a variety of such tests, Horowitz concluded that whether dogs take on a guilty look— lowered gaze, ears pressed back, slumped body, averted head, tail rapidly beating between the legs—is unrelated to them having followed orders or not. It is not about what they have done but about how their owner reacts. If the owner scolds them, they act extremely guilty. If the owner doesn't, everything's fine and dandy. In fact, many transgressions by dogs occur long before the owner comes home, so that the association imprinted on their mind is between owner and trouble, not so much between behavior and trouble. This is why dogs often happily parade the evidence of their misdeeds right in front of you, such as a chewed-up sneaker or a dismembered teddy bear.[28]

The behavior of dogs after a transgression is best regarded, then, not as an expression of guilt but as the typical attitude of the member of a hierarchical species in the presence of a potentially annoyed dominant: a mixture of submission and appeasement that serves to reduce the likelihood of attack. I have only felines at home, no canines, so I never get to see the slightest trace of guilt in my pets, which has to do with the less hierarchical nature of cats. Dogs are sensitive to rule violations and understand them better. The original template of guilt remains the social hierarchy, even though humans internalize fear of punishment to such a degree that we blame ourselves. We essentially castigate ourselves by feeling bad about behavior that we agree we shouldn't have shown, or about behavior that we should've shown but failed to. We are ready for atonement, such as making amends or accepting punishment.

This kind of internalization is rare or absent in other species, but it can't be ruled out. One problem is that we have been overly anthropo-centric in testing our pets, using rules that make perfect sense to us but probably not to them, such as "Don't jump on that couch!" or "Don't put your nails into my leather chair!" Humans come up with the weird-est prohibitions! It must be as hard for animals to grasp the relevance of these rules as it is for me to understand why I can't chew gum in Singapore.

Perhaps we should instead test animals on behavior that is wrong by almost any standard, including that of their own species. Konrad Lorenz gave a perfect example for his dog, Bully, who broke the fundamental rule never to bite your superior. Humans don't need to teach this rule,

Mama, the long-time matriarch of the Burgers Zoo chimpanzee colony, with her daughter Moniek. At the time of this photo, Mama was at the height of her power. She did not physically dominate any full-grown males but nevertheless wielded immense political influence.

At the age of fifty, Mama looked old and walked with great difficulty due to arthritis. Despite this, she was still very much respected.

Mama was known as the best dispute mediator. Here, she intervenes in a squabble between the alpha male, Nikkie (*right*), and her own protégé, a juvenile male named Fons, who screams in protest (*left*). Mama steps between them while pant-grunting at Nikkie, which calms him down. She will proceed to groom him, then lead Fons away.

A juvenile rhesus macaque grins at an approaching dominant male. This expression with pulled back lips and closed mouth indicates submission but also willingness to stay.

The teeth-baring of many primates, including the human smile, is thought to derive from a reflex-like response to noxious stimuli. Here a cactus-eating baboon in Kenya retracts her lips to keep them away from the prickly plant.

A female rhesus monkey aims the typical threat of her species at a subordinate: she fiercely stares at the other while opening her mouth without showing much of her teeth.

Orange, the alpha female of a rhesus monkey troop, sits between her two adult daughters, who have approached her after a vicious fight between them. During this family reconciliation all three females utter a chorus of friendly grunts and lip-smacks while focusing attention on each other's infants.

Body contact is calming for all primates. These two female chimpanzees hold on to each other while watching a tense fight in their community.

During a snowstorm in Jigokudani Park, in Japan, macaques groom in a hot water spring. Primates spend an extraordinary amount of time grooming, which serves to maintain bonds and supportive relationships.

Capuchin monkeys pay close attention to food possessed by others. They share easily but are also sensitive to inequity.

Primates may throw noisy temper tantrums when their expectations are violated. This juvenile (*right*) started screaming when his mother, who carries a baby, pushed him away from her. Until the new birth, he used to cling to her belly.

The most common expression of empathy is consolation, a comforting reaction to others in distress. A bonobo at the Lola ya Bonobo Sanctuary tenderly holds another who has just lost a fight. Photograph courtesy of Zanna Clay.

A female chimpanzee (*right*) kisses the alpha male on the mouth after a fight between them in which the male chased her. As in humans, chimpanzees kissing is typical for reconciliations and for greetings after separation.

I am holding Roosje at Burgers Zoo in 1979. We successfully trained an adoptive mother, Kuif, to bottle-feed this baby chimp. Photograph taken in 1979 courtesy of Desmond Morris.

A juvenile chimpanzee screams in protest while begging with outstretched hand for the berries that an adult male had stolen from him.

Ever since Darwin, debates have raged about whether frowning, caused by little muscles between the eyebrows, is uniquely human. We now know that other primates have the same muscles, which they also contract during anger. A juvenile bonobo male (*left*) stares with a frown and tense mouth at his opponent, a younger male, who has sought the protection of a female (*right*). She has put an arm around him while making hitting movements to deter the aggressor.

Like human smiling, bonobo teeth-baring often serves to appease others and improve their mood. Here Loretta (*right*) solves a standoff with baby Lenore, who kept reaching for her food, a bundle of leafy branches. Loretta's problem is the presence of the baby's dominant mother (*left*). She moves the food out of Lenore's reach while offering her a handshake and friendly grin instead.

Two adult male chimpanzees have ended high up in a tree at the end of a fight. One of them holds out his hand to the other in an invitation to reconcile. Just after I took this picture, the two males embraced and kissed, then climbed down together.

The chimpanzee's loudest vocalization is the scream, which expresses fear and anger. It is typically aimed against high-ranking individuals, such as by these two females angrily chasing an adult male.

Alpha males live under constant pressure and can get stressed out. This male at the Yerkes Field Station had a rival who'd never let up and provoked him every day. The constant worry seemed reflected in the alpha's eyes.

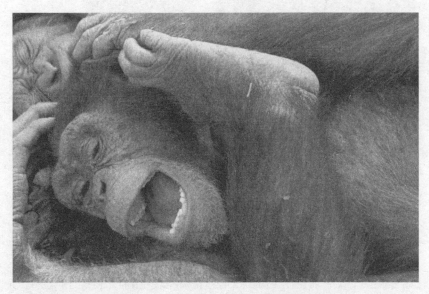

Apes utter hoarse panting laughs during playful roughhousing and chasing.

An adult female (*left*) and adolescent male bonobo (*right*) stand upright at the San Diego Zoo. Of all the great apes, the bonobo is built most like our ancestors, with their relatively long legs, their foot shape, and their brain size. Since bonobos are genetically as close to us as chimpanzees are, they deserve equal attention from anyone interested in human evolution.

and indeed Lorenz notes that Bully had never ever been punished for it for the simple reason that he'd never broken it. But the dog accidentally bit his master's hand when Lorenz tried to break up one of the fiercest dogfights he had ever seen. Even though he did not reprimand Bully and immediately tried to pet him, the dog was deeply upset by what he had done and suffered a complete nervous breakdown. For days afterward, he was virtually paralyzed and ignored his food. He would lie on the rug breathing shallowly, occasionally interrupted by a deep sigh coming from inside his tormented soul. You'd have thought that he had come down with a deadly disease. For weeks, Bully remained subdued. He had violated a natural taboo, which among members of his species, or their ancestors, could have had the worst imaginable consequences, such as expulsion from the pack. We seem to be getting close here to an internalized rule, the violation of which may lead to profound emotional and physical misery that is probably not far removed from guilt.[29]

What about our close relatives, the primates? Do they ever go this far? One of the best-known external regulators in their society is the effect of high-ranking males on the sex life of lower-ranking ones. As a student working with long-tailed macaques, I followed activities in an outdoor section of their group cage connected to the indoor section by a tunnel. Often the alpha male would sit in the tunnel so he could keep an eye on both sides. As soon as he moved indoors, however, other males would approach the females outdoors. Normally, they would be in deep trouble for doing so, but now they could mate undisturbed. Fear of punishment did not entirely disappear, however. They'd regularly run to the tunnel entrance to take a peek inside to check on the alpha, worried about his sudden return. If they encountered him shortly after a sneaky copulation, the low-ranking males would have wide toothy grins on their faces, betraying nervousness, even though the alpha couldn't possibly know what had happened. When this kind of situation was tested systematically in an experiment, the same reactions were observed, leading the investigators to dryly conclude that "animals can incorporate behavioral rules which are associated with their social role and can respond in a manner that acknowledges a perceived violation of the social code."[30]

Social rules are not simply obeyed in the presence of dominants and forgotten in their absence. If they were, low-ranking males wouldn't need to check on the whereabouts of the alpha when he is out of sight, or show exaggerated submission to him after forbidden exploits. They internalize rules to some degree. A more complex expression occurred once in the Arnhem chimpanzee colony following the first time the beta male, Luit, defeated the alpha, Yeroen. The fight had occurred while both males were alone in their night quarters. The next morning the colony was released onto its island, only to discover the shocking physical consequences of the fight:

> *When Mama discovered Yeroen's wounds she began to hoot and look around in every direction. At this, Luit broke down, screaming and yelping, whereupon all the other apes came over to see what was the matter. While the apes were crowding around him and hooting, the "culprit," Luit, also began to scream. He ran nervously from one female to the next, embraced them, and presented his behind to them. He then spent a large part of the day tending Yeroen's wounds. Yeroen had a gash in his foot and two wounds in his side, caused by Luit's powerful canines.*[31]

Luit's situation resembled that of Bully the dog, in that he had broken the spell of the hierarchy. Over the preceding years, no one had ever injured Yeroen. *What a terrible thing to do!* the group's reaction seemed to convey. Luit did his best to make amends, but not by giving up his strategy to dominate Yeroen, because in the weeks that followed he kept the pressure on, and in the end he forced Yeroen into retiring from the top spot. Had Luit reacted to the injuries because he felt guilty based on an internal rule of how he ought to behave? Or had he just been worried about how the colony might react?

Here bonobos go beyond chimpanzees. Violence is so rare in this species that it troubles them more. Attackers seem to mix regret about their actions with empathy, because they hurry to make up afterward. By contrast, in other primates, reconciliation is more typically initiated by the subordinate. Bonobos are different in that it is usually the dominant individuals who seem remorseful, especially if they have caused injuries. I remember one of them returning to his victim and without hesitation reaching for the exact toe that he had bitten so as to examine the damage. His behavior indicated that he knew precisely what he had done and also where. For me, if there is one situation that hints at remorse, it is these scenes of dominant bonobos running back to their victims to spend half an hour or more licking and cleaning the wounds that they themselves had inflicted.

It is hard to be sure what bonobos feel, but I must add that in my more cynical moments, I ask very similar questions about human guilt. Aren't we overrating the power of internalization? Look how people throw all inhibitions overboard when conditions change, such as during war or famine or political unrest. Many an upstanding citizen loots, steals, and kills without compunction when there is little chance of them getting caught or when resources have become scarce. Even a less dramatic change of circumstances, such as a vacation in a faraway land, may prompt people to act in outrageous ways (public intoxication, sexual harassment) unthinkable in their hometown.

People who say they feel guilty and apologize for their misdeeds do not always convince me either. In fact, I prefer silent guilt. Apologies by public figures are so full of phony emotions and false tears that they have become known as "nonpology" or "fauxpology," defined as a statement in the form of an apology without the actual acceptance of responsibility. Often, the victim is blamed, such as in *I am sorry you felt offended by my tweet*. In 1988, Jimmy Swaggart, a prominent American televangelist, was caught with a prostitute. Afterward he cried and cried on television, his face turning into a river, begging God and his congregation to forgive his sins. A few years later, he got caught again. What passes for an expression of guilt in humans is often, just as in dogs, a way to avert negative consequences rather than evidence of a deeply felt distinction between right and wrong.

I am not denying that humans can make this distinction, or are capable of actual guilt, but the difference between that and appeasement/submission is not nearly as sharp as we'd like to think.[32] Guilt is often described as the product of religion and culture, or framed as an emotion that urges us to make amends and repair the harm we've done. This is all very nice and undoubtedly true, but we shouldn't downplay the fear factor. Guilt and anxiety often go together, feeding on each other. Underneath it all is something much more fundamental than culture or religion. What feeds guilt and shame is a deep desire to belong, a survival issue for any social animal. The greatest underlying worry is rejection by the group. This is what drove Luit to embrace the females gathered around his wounded rival, Bully to go into a depression, teenagers to be embarrassed by their parents, and Swaggart to shed his crocodile tears. Concern about upsetting others and losing their love and respect is what ultimately drives human shame and guilt.

Since this fear underlies similar behavior in other species, let me close with a description of the way Gua—a young chimpanzee raised in the home of Winthrop and Luella Kellogg in the 1930s—typically responded to the rebukes of her adoptive parents. I don't necessarily regard Gua's reaction as proof of shame or guilt, but she does exhibit a deep desire to belong and be forgiven, which for me is at the root of both emotions. The Kelloggs described how, if things ended well, Gua would invariably give a deep sigh of relief:

> *When Gua was punished, or often simply scolded, for biting the wall, making an evacuation error, or committing some similar faux pas, she would utter "oo-oo" cries and try to run to our arms. [If we pushed her away] this would invariably precipitate more severe outbursts of wailing and screaming, which would only subside when we signaled our willing-*

ness to receive her. The vocalizations would then change to a very rapid rhythm of "oo-ooing" and she would at the same time rush towards us with arms outstretched. She would pull herself upward to our shoulders—at all odds getting her face somewhere near our own. The next reaction was the kiss of reconciliation. If we acquiesced and responded to her invitation she would then heave her great sigh, audible a meter or more away.[33]

The Yuck! Factor

The nose wrinkle of disgust is typical on a rainy day. I call it the chimpanzee's "rain face." As soon as a downpour starts, all chimps, young and old, pull an ugly face, pulling their upper lip close to their nose and sticking their lower lip slightly out. Their eyes are semiclosed, their teeth visible. Chimps hate to get their hands wet, so they show this face while walking bipedally through the wet grass with their hands neatly folded in front of their chest, looking utterly miserable. I am familiar with the same face in humans, because the Netherlands is the number-one biking country. Thousands of cyclists crisscross the big cities in thick throngs, rain or shine, because this is how they get to work or school. Whenever it rains, cyclists show rain faces inside their plastic ponchos, bothered by the weather and the prospect of wet clothes for the rest of the day.

Disgust and dislike are among the oldest emotions and among the few to be linked to a specific brain area: the insular cortex (also known as the insula). Activation of this area creates a strong disgust for anything inside one's mouth. Thus a monkey chewing peanuts with gusto will nevertheless spit them out as soon as his insula is stimulated. At the same time, he will change his facial expression, wrinkling his upper lip and nose together while using his tongue to shove food out of his mouth.[34] In human subjects, the same insula lights up when they view pictures of things that make them gag, such as excrement, rotting garbage, or maggot-infested food. We, too, pull the upper lip closer to the nose while narrowing our eyes and knitting our eyebrows. The characteristic nose wrinkle is a ritualization of muscle contractions that protect the eyes and nostrils against incoming danger, such as wafts of foul air. In English, we say we "turn up our nose" at something.

The facial similarity and the engagement of the same brain area in monkeys, apes, and humans hint that it's all the same emotion. Disgust is older than the primates, though, because all organisms need to reject dangerous substances and parasites. Rats open their mouth wide (called "gaping," probably a vomiting intention movement) when smelling foods that make them nauseous. Cats recoil from perfumes or madly shake their paw after touching a sticky surface. Dogs whine and pucker in reaction to sour smells. Endearingly, when cats encounter a malodorous object, such as a dead cockroach, they scrape around the filthy thing with a paw to cover it, even if there is no dirt around, such as on the kitchen floor. All these reactions boil down to self-protection against harmful substances. "Visceral disgust," as it's known, is a behavioral extension of the immune system, coming from deep inside and nearly impossible to control.

In a curious twist, disgust has become the Cinderella of the emotions. Despite its humble beginnings, no feeling nowadays gets more love and attention from psychologists, because of its connection to morality. We feel disgust at certain kinds of behavior, such as incest and bestiality but also political corruption, treason, fraud, and hypocrisy. Appalled by selfish people who fake cancer to invite donations on the Internet or who take a parking spot they're not entitled to, we call them "disgusting" and say they "leave a bad taste in the mouth." Whenever politicians want to turn us against people in our midst, such as certain ethnic groups, they play the disgust card. They say these people resemble certain animals that we detest, or smell like them. They even pull the wrinkle-nosed face while discussing them. Conversely, we equate cleanliness with virtue and good things. When we "wash our hands" of a shady affair, we are saying, like Pontius Pilate, that purity equals innocence.[35] The current literature on "moral disgust" sometimes goes overboard by treating the original emotion almost as an afterthought. It elevates human disgust to a cultural phenomenon, an acquired taste far removed from mere avoidance of pathogens.

We commonly call foods that repulse us "disgusting." We learn our eating habits from others in our own culture, so we may strongly dislike even the favorite food of another culture. Once in a bar in Sapporo, I got a standing ovation as the first Westerner—or so they told me—to eat half a bowl of *natto*, a smelly fermented soybean dish. I felt rather proud, but then someone asked me how I liked the stuff. Before I could come up with a diplomatic answer, my face betrayed my feelings. Everyone laughed. The Japanese, on the other hand, can't stand the skins of apples and pears, which they always peel, something I find odd. Clearly, we humans have acquired tastes and acquired disgusts. Animals make no such cultural distinctions, or so the argument goes, because they instinctively know what to eat and what not.

Another popular idea is that disgust helps us set ourselves apart from animals by classifying their bodies and products as revolting. Rotting plants and fruits don't bother us nearly as much as the decaying corpses of animals and their feces, blood, semen, intestines, and so on. And it's not just the sight and smell of dead animals that sicken us, the theory goes; they remind us of our own mortality. We are so scared of death that we loathe everything that stresses our commonality with animals and their fragile existence. Recoiling from animals helps us deal with existential questions, which explains why some scientists consider disgust nothing less than a sign of civilization!

My head is spinning from these highfalutin notions about a straightforward emotion that evolved to keep dangerous substances at bay. Academics tend to get carried away by their own fancy elaborations. So effectively have they managed to muddle the tracks leading back to disgust's mundane origin that it has begun to look like a brand-new emotion. And not just an emotion—it's viewed as a mental operation that defines us and explains our noblest achievements. Not all psychologists think this way, though. Some feel, like me, that if we look deep into the feeling of disgust, even when applied to the moral domain, we find the exact same emotion located in the insula and expressed with a wrinkly nose.

What gets me especially, as someone who loves animals and works with them every day, is the thought that loathing them somehow advances civilization. If so, why do we massively bring animals into our homes, where we pamper them like family despite the pee and poop we regularly have to clean up? Cat lovers are undeterred by the litter box, and dog lovers by the poop scooper, not to mention what horse lovers have to deal with. Look at how dependent humanity is on animals. We use (or have used) them not only for food but also to plow land, carry armies, send airmail (pigeon post), smell out drugs, assist with hunting, herd sheep, console the sick, catch rodents, pollinate flowers, and so on. If humans are revolted by animals, why do zoos attract an estimated 175 million visitors per year in the United States alone? Think of all the animal videos everyone watches on Facebook. Animated movies for kids are full of talking animals. Toy stores sell plush bears, elephants, dinosaurs, and so on, which our children clutch while they sleep. Humans are in fact extremely attracted to animals and celebrate them with expressions such as "brave as a lion," "wise as an owl," and "busy as a beaver." While it is true that in the West we like to view ourselves as separate from the animal kingdom, our ancestors, who lived closer to nature, did not likely entertain the same illusion. They probably worshipped animal gods the way preliterate peoples still do. I don't think, therefore, that human disgust has anything to do with the denial of our animality.

That this emotion is thought to have a cultural origin is intriguing in light of what we know about the cultures of other species. It is very well possible that animals, too, have cultural disgusts. Perhaps some species know instinctively what to eat, such as those that rely on a single food— giant pandas munch bamboo all day and koala bears ingest only eucalyptus leaves—but this is a rare situation. The tropical rain forest contains thousands of different plants, of which primates eat the fruits and leaves. The majority of these plants are inedible, some are poisonous, and others make them sick, so how do primates know which species to exploit? They are very careful about what they eat and at what stage of ripeness. In fact, color vision is thought to have evolved in the primate order to assist with these distinctions. Chimpanzees also eat substantial amounts of meat, which they hunt themselves. They must have the same sensitivity to decaying carcasses that we do, because they pass up opportunities to scavenge dead animals that they have not killed themselves. This aversion explains why Tara's game with the rat cadavers worked so well.

We know from a wealth of research that young chimps learn from their elders not only what to eat and what to avoid but also how to access hard-to-reach foods. They learn how to fish for termites, crack nuts, and collect honey from beehives. Our own experiments demonstrate that apes are excellent imitators,[36] which in their natural habitat translates into culturally acquired food preferences. Nowadays culture studies cover a wide range of species, from birds and fish to dolphins and monkeys. How this bears on disgust can be illustrated with an elegant field experiment in a South African game reserve.

The Dutch primatologist Erica van de Waal (no relation) gave wild vervet monkeys open plastic boxes filled with maize corn, a food that these small grayish monkeys with black faces love. But there was a catch. Some of the kernels were blue, and some were pink. For one group, the blue kernels were good to eat, whereas the pink ones were laced with aloe, which tastes disgusting. For another group, treatment was reversed: the blue ones were treated with aloe, while the pink ones tasted good. Depending on which color corn was palatable and which not, some monkeys learned to eat blue, and others pink, by associative learning. But then the investigators removed the distasteful treatment from all the kernels and waited for new infants to be born. Several groups of monkeys now received perfectly palatable corn of both colors, but they all stubbornly stuck to their acquired preference, never discovering the improved

taste of the alternative color. Of twenty-seven newborn infants, only one
learned to eat food of both colors. The rest, like their mothers, never
touched the other color even though it was freely available and tasted just
as good as the other. Some youngsters even sat on the edge of the box
with the rejected corn while happily feeding on the preferred kind. The
single exception was an infant whose mother was so low in rank, and
so hungry, that she occasionally tasted the forbidden fruits. So even this
infant followed her mother's feeding habits.[37]

The power of conformism is immense. Far from being an extrava-
gance, it is a widespread practice. By following their mothers' example of
what to eat and what to avoid, infants stand a better chance in life than
by trying to figure out everything on their own and running the danger of being poisoned. The implication, of course, is that animals, too,
may acquire disgust. Adult monkeys came to reject the unpalatable corn,
then transmitted their preference to their offspring. Whether the infants
were actually disgusted by the corn their mothers refused to eat is hard
to tell, but behaviorally speaking they showed a clear attraction to one
type and an aversion to the other, which in humans we wouldn't hesitate
to put in emotional terms.

On the subtropical island of Koshima, in Japan, the French primatologist Cécile Sarabian studied the yuck! response of wild macaques. She
placed three different items near each other on the beach: monkey feces,
realistic-looking plastic feces, and a brown plastic notebook cover. On
each item, she then placed either a grain of wheat (which monkeys eat but
not avidly) or half a peanut (which monkeys love). Upon discovering these
items, monkeys picked off and ate all the peanuts (although they sometimes vigorously rubbed their hands after having touched the feces). They
also picked off all the wheat grains from the plastic notebook cover but
only about half the grains from the real and fake feces. That is, the monkeys were disgusted enough by the poop piles to forgo the grains, whereas
when it came to peanuts, their desire won out. Feeding on potentially con-
taminated foods is always a balancing act between aversion and nutritional

value, which is higher for the peanuts. Sarabian is now applying similar tests to chimpanzees and bonobos, finding that some contaminants are disgusting enough for them to reject perfectly fine foods.[38]

Disgust may also be triggered by impurities and dirt when there is nothing to eat. Rain isn't even filthy, but like our fellow apes, we dislike it enough to pull a face. A filthy taxicab interior disgusts us, as does the messy bathroom of other people. Just as we shower and brush our teeth in the morning because we care about our well-being (the functional side) and because we hate being filthy (the emotional side), animals pursue bodily hygiene not just for its health benefits but also out of a desire for cleanliness and a deep aversion to impurities. Look at how meticulously a preening bird cleans itself with its beak, especially the pennaceous feathers (long stiff flight feathers) on their wings and tail. It is hard not to admire their hygiene.

Moreover, it makes them happy. Back in my student days, once a week I would let my tame jackdaws spray my dorm room by splashing around in a large water basin placed on the floor. For the rest of the morning, they'd preen every single feather on their body. By the end, they'd be all fluffed up and would burst into "song" (between quotation marks, because jackdaws don't make very pleasant sounds), obviously in an excellent mood by their immaculate state. The same fastidiousness can be seen in cats carefully washing their face and every other part of their body. In stalking animals, cleanliness helps hide their scent from prey. House cats are said to spend as much as 25 percent of their waking time grooming themselves to reach a spotless condition.

A desire for neatness and cleanliness outside the body is common in many species. Nesting animals typically prefer order and a debris-free environment. The male bowerbird places hundreds of small decorations (flowers, beetle wingcases, shells) in the court of his bower to attract females—and constantly arranges and rearranges them. Songbirds fastidiously remove the fecal sacs (feces inside a mucous membrane) that their chicks eliminate: they take the white sac in their beak and fly off to drop it away from their nest. Naked mole rats have special toilet chambers in their tunnel system—when the old ones get dirty, they plug them up with soil, while digging fresh ones elsewhere. The advantages of cleanliness are clear enough—clean feathers promote flight and body isolation; a clean nest keeps parasites and predators away—but we need to pay more attention to the underlying emotions, which likely include a strong aversion to anything that doesn't belong. Disgust at impurities marks tons of species.

Finally, disgust among animals can also be social in nature, as psychologists are so fond of discussing among humans. Other primates can indeed be turned off by social acts or by certain individuals. The first example that jumps to my mind is an anecdote about Washoe, a chimpanzee trained in American Sign Language. She had learned to sign "dirty" for soiled furniture and clothes. Once when a macaque was seriously annoying her, she repeatedly signed "Dirty monkey! Dirty monkey!" This was a novel use of the word and not one she had ever been taught. It suggests that to Washoe, social aversion felt like dealing with filth.

Disgust at individuals has a sexual context when older males court females. I have watched adolescent female chimps literally run off screaming when an old male tried to have sex with them. During the mating season, rhesus monkey females, too, sometimes leave the area as soon as they see an older male heading in their direction. The young females are probably trying to avoid fertilization by males old enough to be their fathers—hence incest avoidance—but they surely act as if they are horror-struck. When the approaching male is their own kin, their revulsion is even more obvious. One wild chimpanzee refused her own son's sexual advances but eventually submitted when he kept intimidating her. She did so under protest, though, and "screamed loudly throughout and jumped away prior to ejaculation."[39]

In the 1960s, Gombe National Park experienced an outbreak of polio. As Jane Goodall describes, the afflicted chimpanzees suffered paralyzed limbs that left them unable to move through the forest or climb trees.

They were forced into bizarre patterns of locomotion. Healthy chimps in the community were extremely disturbed by their presence. They might approach them but then would stop a safe distance away, sometimes with soft *hoo* alarm calls. They rarely touched an afflicted member and never groomed them, which is highly unusual. Despite his nonfunctioning legs, one adult male made an extraordinary effort to join two grooming males in a tree, but both of them kept moving away, leaving him without company.[40]

Even animal disgust at excrement has a social component. Mother apes commonly get soiled by their youngest offspring, which they carry around the whole day. They take this very calmly, like part of the job. They usually detect from their infant's behavior that an elimination is coming and hold the infant slightly away from their body. If they are too late, they just pick up a few leaves and clean off the mess. In contrast, if a chimpanzee is attacking another, whose fear diarrhea then soils the attacker, the latter will rub it off with desperate frantic movements, clearly upset by the unexpected staining. It's not just the excrement that bothers them but its source.

Disgust reactions to members of an out group are even stronger and may extend to inanimate objects linked to them. If during a patrol of their borders, chimpanzee males spot a night nest built by neighboring males on their side of the forest, they naturally take this as an insult. Several males will climb up the tree and carefully smell and inspect the nest, after which they shake it apart and tear out every single branch until the nest is destroyed. I imagine that a dog who finds his enemy's marking on his territory, and deliberately urinates over it, is driven by the same kind of revulsion. A funny story along these lines: one night a fieldworker on the African plains placed his boots outside his tent. The next morning he felt something squishy inside a boot, which turned out to be leopard droppings. The big cat must have deemed the smell of his feet offensive and decided to obliterate it.

There aren't many animal examples of disgust triggered by the *behav-*

ior of others equivalent to what we call moral disgust. This doesn't mean
that they don't occur, though. No one has been looking for them, except
in a handful of studies on how primates evaluate the "character" of others. Scientists at the University of Kyoto tested how capuchin monkeys
reacted to a scene in which a person pretended to have trouble opening a plastic container and asked a human experimenter for help. The
experimenter kindly gave the help. In the next scene, the person asked
a different experimenter for help—one who turned away and ignored
the request. Would the monkeys like the good guy or the selfish jerk?
Mind you, this was about how the experimenter treated not the monkeys but another person. After watching the scenes enacted in front of
them, the monkeys refused to have anything to do with the despicable
experimenter, turned off by her poor level of cooperation.[41]

Such experiments, which are being conducted more and more often,
bear on the evolution of morality. This is a topic close to my heart, one
I've treated in previous books. Of the many illustrations that I could
offer, let me highlight just one story concerning a social norm violation
among chimps. It happened when Jimoh, the alpha male of the colony
at the Yerkes Field Station, suspected that an adolescent male and one of
his favorite females were secretly mating.

I followed the whole affair from my office window, which allowed
me to see every corner of the compound. For the chimps on the ground,
however, the view was obstructed by many obstacles. This enabled the
young male and female to temporarily escape from Jimoh's view. The
alpha male realized that something was up, however, and went looking for them. Normally, he would merely chase off the culprit, but for
some reason—perhaps because the same female had refused him earlier
that day—he went full tilt after the young male and did not relent. I
should add that even though adult males often slap youngsters around
or roughly trample them, the females in this colony didn't allow them
to sink their canines into them—they drew the line there. Jimoh was
mad, however, and pursued the young male all around the enclosure,

bringing about total panic in his victim. He seemed intent on catching and punishing him.

Before he could reach this point, however, females close to the scene began to *woaow* bark. This indignant sound is a warning call against aggressors and intruders. At first, the callers looked around to see how everyone else was going to respond. Others joined in, especially the alpha female, and the intensity of their calls increased until literally everyone's voice was heard in a deafening chorus. It gave the impression of the group taking a vote. Once the protest swelled to a crescendo, Jimoh broke off his attack with a wide nervous grin on his face: he got the message.

I felt like I was seeing moral disapproval at work.

Emotions Are Like Organs

Let me start off with a radical proposal: emotions are like organs. They are all needed, and we share them all with other mammals.

When it comes to organs, this is obvious. No one would argue that some organs are fundamental, such as the heart, brain, and lungs, whereas others are less needed, such as the pancreas and kidneys. Anyone who has ever had a problem with their pancreas or kidneys knows that each and every organ in our body is indispensable. Moreover, our organs don't fundamentally differ from those of rats, monkeys, dogs, and other mammals. But I wouldn't even limit it to mammals. Apart from mammary glands, which set *Mammalia* apart, all organs are shared across the vertebrates, including frogs and birds. I dissected many a frog as a student, and they have everything, including reproductive organs, kidneys, a liver, heart, and so on. The vertebrate body requires a certain machinery, and if any part is missing or failing, it dies.

Regarding the emotions, though, the thinking is patently different. Humans are thought to have only a few "basic" or "primary" emotions, which are essential for survival. The number varies by scientist, running anywhere from two to eighteen, but usually it's around half a dozen. The obvious basic emotions are fear and anger, but there's also arrogance, courage, and contempt. The idea that some emotions are more basic than others goes back to Aristotle and has been elevated to a theory, known as Basic Emotion Theory (BET). For an emotion to be counted as "basic," it must be expressed and recognized by humans all over the world and be hard-wired—a way of calling it inborn. Basic emotions are biologically primitive and shared with other species.[42]

Human emotions that lack stereotypical expressions, on the other hand, are known as "secondary" or even "tertiary." They enrich our lives, but we can manage without them—we'd still be fine. They are moreover entirely our own and vary by culture. The list of proposed secondary emotions is quite lengthy, but as you will have noticed, I take issue with the entire proposal that they even exist. It's as flawed as would be the idea that not every organ in our body is essential. Even the appendix (the blind-ended little tube connected to the cecum) is not called "redundant" or "vestigial" anymore, because it has evolved independently so many times that its survival value is not in doubt. Its probable function is to harbor good bacteria, which help reboot the digestive tract, for example, after a severe case of cholera or dysentery. In the same way that every part of our body has its purpose, every emotion evolved for a reason.

First, as we have seen here for pride, shame, guilt, revenge, gratefulness, forgiveness, hope, and disgust, we can't exclude their presence in other species. These emotions may be more developed in us, or they may be used under a wider range of circumstances, but they aren't fundamentally new. That some human cultures emphasize some of them more than others hardly argues against a biological origin.

Second, it is highly unlikely that any common emotion is functionless. Given the cost of getting all worked up and passionate about something, and given how much such states affect decision making, superfluous emotions would pose an incredible burden. They might lead us astray, which is certainly not the sort of baggage natural selection would let us carry. Hence my proposal that all emotions are both biological and essential. None is more basic than the others, and none are uniquely human. To me, this is a logical position given how closely the emotions are tied to the body and how all mammalian bodies are fundamentally the same. Thus, when human subjects were asked to guess the state of emotional arousal of a variety of reptiles, mammals, amphibians, and other land animals, just from listening to their calls, they were remarkably good at doing so. There seem to exist "acoustic universals" that allow all vertebrates to communicate emotions in similar ways.[43]

Note that I am not talking about feelings here, which are harder to know than emotions and may be more variable. As subjective evaluations of one's own emotions, feelings may well vary from culture to culture. What animals feel is hard to know, but it is good to realize that the inaccessibility of their feelings has two sides: we can only guess what they feel, but we also can't rule out any particular feelings. Given how often the second warning is ignored, let me briefly return to the standard way of dismissing animal feelings, which is to shift the attention to behavioral functions and outcomes. As soon as you propose that two animals love each other, you'll hear that they don't need to because all that matters is reproduction. If you propose that they show pride, you'll hear that they just enlarge themselves to show off. If you propose that an animal is afraid, you'll hear that animals don't need fear so long as they manage to escape danger. It always comes down to the outcome of behavior.

This is a rather underhanded maneuver, though, because beneficial outcomes never exclude emotions. In biology, this is known as the confusion between levels of analysis, against which we warn our students every day. Emotions belong to the motivation behind behavior, whereas outcomes belong to its functions. The two go hand in hand: every behavior is marked by both motivations and functions. We humans both love *and* reproduce, feel pride *and* intimidate, get thirsty *and* drink water, are afraid *and* protect ourselves, and are disgusted *and* clean ourselves. So stressing the functional side of animal behavior doesn't in any way get us around the question of emotions—it merely sidesteps it.

Think of this next time someone claims that animals have sex just to propagate. That can't be the whole story. Members of the opposite sex still have to come together, be drawn to each other, trust each other, and get aroused. Every behavior has its mechanism, which is where the emotions come in. Copulation requires the right hormonal conditions, sexual desire, mate preference, compatibility, even love. This is as true for animals as it is for us.

Curiously, love and attachment are rarely listed as basic human emotions, but they strike me as essential for all social animals, and not only in the context of sex. We find strong, lifelong pair-bonds in many birds and some mammals, bonds that endure regardless of sex (including long seasons without any). The mother-offspring bond is typical of mammals and may cause deep distress when a mother loses an infant. It is impossible to look at a mother ape playing with her child, lifting it into the air while gently turning it around (known to us as the airplane game), nor at elephant mothers and aunties displaying extreme alertness to their calves, and not see love. The only reason love isn't classified as a basic emotion is that it fails to show up in the face. We don't have a love facial expression, the way we have expressions for anger and disgust. To me, this shows the limitation of the traditional focus on faces, which is felt even more keenly in relation to the many animals that lack facial flexibility.

The never-ending debate about how to classify emotions, or even what an emotion is, reminds me of a phase in biology when our main preoccupation was the classification of plants and animals. This field, known as *systematics*, had its heyday in the eighteenth and nineteenth centuries. There are few debates more heated (or more fruitless) than those about whether a species deserves to be a species on its own or is better considered a subspecies. DNA is settling many of these disputes, the way that neuroscience will probably assist the classification of emotions. If two emotions, such as guilt and shame, share activations in the brain and are expressed in similar ways, they obviously belong together. They are like two subspecies of the same self-evaluative emotion, even though—like any good naturalist—we love to dwell on their distinction. On the other hand, emotions, such as joy and anger, that share few brain activations and bodily expressions belong on divergent branches of the emotion tree. While not everyone is convinced that every emotion has its own neural signature, plotting out all the brain areas and circuits involved is our best bet for building an objective taxonomy of emotions, one based on hard science, the way we use DNA comparisons to map taxonomies of animal and plant families.

Neuroscience may also help determine which emotions are homologous across species. We know already of brain similarities between dogs and businessmen anticipating a reward, and our next step might be to put a "guilty" dog in the fMRI scanner to determine if the same brain circuits are active as in human subjects asked to imagine guilt.

This brings me back to the insula and its role in disgust at unpalatable foods, at immoral behavior, and as in the case of the Gombe chimpanzees, at those afflicted by disease. Instead of viewing every kind of disgust as a separate emotion, why couldn't they all be the same? The triggers of disgust vary with species, conditions, and even cultures, but the emotion itself, and perhaps also its associated feelings, involves a shared neural substrate. Robert Sapolsky, an American primatologist and neuroscientist, in a funny first-person voice, describes how evolution may have produced moral disgust by strapping it onto an existing emotion:

> Hmmm, extreme negative affect elicited by violations of shared behavioral norms. Let's see . . . Who has any pertinent experience? I know, the insula! It does extreme negative sensory stimuli—that's like, all it does—so let's expand its portfolio to include this moral disgust business. That'll work. Hand me a shoehorn and some duct tape.[44]

This could be the story of all human emotions: they are variations on ancient emotions that we share with other mammals. Darwin defined evolution as descent with modification, which is another way of saying that evolution rarely creates anything completely new. All evolution does is refurbish old traits, turning them into ones that suit current needs. This is why none of our emotions are entirely new, and they all play an essential part in our lives.

5 | WILL TO POWER

Politics, Murder, Warfare

In July 2017, when Sean Spicer, then the White House press secretary, was discovered hiding in the bushes to dodge questions from reporters, I knew Washington politics had become truly primatological. A few weeks earlier James Comey had on purpose worn a blue suit while standing in the back of a room with blue curtains so as to blend in. The tall FBI director hoped to go unnoticed and avoid a presidential hug. The tactic failed.

Making creative use of the environment is primate politics at its best, as is the role of body language, such as sitting on a throne high above the groveling masses, descending into their midst with an escalator, or raising one's arm so that underlings can kiss your armpit, a pheromonal ritual invented by Saddam Hussein. The link between high evaluations of debate performances and the candidates' height is well known—taller candidates have a leg up. This advantage explains why short leaders bring along boxes to stand on during group photos. When Nicolas Sarkozy, the French president, visited a factory, he shipped in a busload of people shorter than himself so that he could stand among them during a photo opportunity. There is no lack of examples, but my list grew exponentially with the last U.S. presidential election, in 2016, when Donald Trump entered the scene.

Like an Alpha Male

Trump's bullying skills against his male rivals were legendary. During the Republican primary, the Donald squashed all his poor fellow candidates by puffing himself up, lowering his voice, and insulting them with demeaning nicknames, such as "Low-energy Jeb" and "Little Marco." Strutting like a male chimp on steroids, he turned the primary essentially into a hypermasculine body language contest. The political issues of the day were secondary. We even heard anatomical comparisons based on the assumption that hand size says something about other body parts. At some unimaginable point in American history, the front-runner held up his hands and asked his audience if they looked small. He guaranteed that the rest of his body was of similar size.

One of Trump's most brilliant moves occurred in response to criticism by Mitt Romney, the 2012 Republican nominee. Trump blasted Romney, reminding his audience that four years earlier Romney had courted him: "You can see how loyal he was, he was begging for my endorsement. I could have said, 'Drop to your knees!' and he would have dropped to his knees."[1] In one swoop, Trump depicted Romney as untrustworthy while creating the visual of him in a prostrate position similar to how a low-ranking chimpanzee crawls in the dust for an alpha.

But even though Trump had intimidation down to a T, this didn't necessarily help him against his female opponent in the general election. Between the sexes, all bets are off. Fighting behavior is bound by rules. Animals capable of killing each other—predators, venomous snakes, horned ungulates—follow standards of engagement. Instead of going all out, they go through ritual motions that test strength and agility without necessarily knocking each other out. In all this, the rules for male-to-male and male-to-female combat are dramatically different, because for a male to kill another male is one thing, but to kill a female is plain stupid. Evolutionarily speaking, the whole reason why a male would try to rise to the top is to have females to produce offspring with. Even though in our political system women vote and are able to occupy the highest office, thus allowing for a social order quite different from that of many other species, the fighting rules have hardly changed. They evolved over millions of years and are far too ingrained to be thrown out. A male generally curbs his physical power while confronting a female. This is as true for horses and lions as it is for apes and humans. These inhibitions reside so deeply in our psychology that we react strongly to violations. In the movies, for example, it's not terribly upsetting to see a woman slap a man's face, but we cringe at the reverse.

This was Trump's dilemma: he was up against an opponent whom he could not defeat the way he could defeat another male. Having watched every presidential debate since Ronald Reagan, I have never seen as odd a spectacle as the second televised debate between Trump and Hillary Clinton on October 9, 2016. Its blatant physicality and hostility made it the debate from hell. Trump's body language was that of a tormented soul ready to punch out his opponent yet aware that if he laid one finger on her, his candidacy would be over. Like a large balloon, he drifted right behind Clinton, impatiently pacing back and forth or firmly gripping his chair. Concerned television viewers live-tweeted warnings to Clinton like "Look behind you!" Clinton herself later commented that her "skin crawled" when Trump was literally breathing down her neck.

Trump's demeanor was of barely contained anger, complete with an actual threat: he said that under his presidency a special prosecutor would throw Clinton in jail. Had he been a male chimp, he would have hurled that chair through the air or lashed out at an innocent bystander to demonstrate his superior strength. Trump did the next best thing, throwing his own running mate under the bus (abandoning him on a foreign policy question) and criticizing President Obama as well as Clinton's husband. He clearly felt more at ease with male targets. In fact, before the debate began, he had held a news conference where he had trotted out several women with accusations against Bill Clinton. None of this solved his dilemma, though, of how to handle a political rival of the opposite gender.

Immediately after the debate, which Trump lost according to most commentators, the British politician Nigel Farage mimicked a feeble version of a chest beat while gushing that Trump had acted like "a silverback gorilla." Right away we got the primate parallels, also reflected in the observations from body language experts. The assumption here was that in order to be an alpha, one must be big and strong, ready to annihilate one's rivals. I have never heard alpha references thrown around as

The term *alpha male* comes from wolf research, where it simply means the top-ranking male. Following Darwin's antithesis principle, dominant (*right*) and subordinate (*left*) wolves adopt opposite postures. The dominant has his hair up and his ears forward; he walks on high legs while growling at the subordinate, who is ready to roll over, has his ears back, and utters high-pitched yelps.

freely as during this period, such as when Trump's son Eric excused his father's lewd banter about women as conversation typical of "alpha personalities." Given that the term *alpha male* quickly rose in popularity after Speaker of the House Newt Gingrich recommended my 1982 book *Chimpanzee Politics: Power and Sex Among Apes* to freshmen congressmen, I feel a need to explain what being an alpha exactly means.

In animal research, the alpha male is simply the top-ranking male of a group. The term dates back to wolf studies of the 1940s by the Swiss ethologist Rudolf Schenkel, and it remains in use. In political parlance, however, it has come to denote a certain type of personality. Ever more business tutorials instruct us on how to become an alpha, emphasizing self-confidence, swagger, and purpose. Alphas are not just winners, it is argued—they beat the hell out of everyone around them and remind them every day who won. They don't let up. A true alpha goes it alone and crushes the competition, like a lion among sheep. These tutorials promote a cardboard version of the whole concept, however, not just for human society but also in relation to wolves and chimpanzees. Alpha males are not born, and they don't achieve their position based purely on size and temperament. The primate alpha male is a much more complex and responsible being than a bully.

Merciless tyrants do sometimes rise to the top in a chimpanzee community, but the more typical alphas that I have known were quite the opposite. Males in this position are not necessarily the biggest, strongest, meanest ones around, since they often reach the top with the assistance of others. In fact, the smallest male may become alpha if he has the right supporters. Most alpha males protect the underdog, keep the peace, and reassure those who are distressed. Analyzing all instances in which one individual hugs another who has lost a fight, we found that although females generally console others more often than do males, there is one striking exception: the alpha male. This male acts as the healer-in-chief, comforting others in agony more than anyone else in the community. As soon as a fight erupts among its members, everyone turns to him to see how he is going to handle it. He is the final arbiter, intent on restoring harmony. He will impressively stand between screaming parties, with his arms raised, until things calm down.

This is where Trump deviated dramatically from a true alpha male. He struggled with empathy. Instead of uniting and stabilizing the nation or expressing sympathy for suppressed or suffering parties, he kindled the flames of discord. It began with him making fun of a disabled journalist and has continued with his implicit support for white supremacists. For the primatologist, the comparisons of Trump's behavior with that of alpha primates are limited, therefore, applying more to his climb to the top than to the execution of leadership.

Meanwhile, Trump kept up his physical intimidations by engaging in white-knuckled handshakes with various world leaders, including younger ones (such as Emmanuel Macron of France), who naturally have a firmer grip than an old man like Trump. Amid these awkward skirmishes, I sometimes wished that bodybuilder-turned-politician Arnold Schwarzenegger would have been allowed to run as a candidate. He is the only one who might have mocked Trump right back with matching physical vigor, perhaps throwing his favorite "girly man" insult around, turning politics into even more of a primal spectacle than it already is.

Political Tantrums

When Aristotle labeled our species a *zoon politikon,* "political animal," he linked this idea to our mental capacities. That we are social animals is not so special, he said (referring to bees and cranes), but our community life is different thanks to human rationality and our ability to tell right from wrong. While he was partly right, the Greek philosopher may have overlooked the intensely emotional side of human politics. Rationality is often hard to find, and facts matter far less than we think. Politics is all about fears and hopes, the character of leaders, and the feelings they evoke. Fear-mongering is a great way to distract from the issues at hand.

Even the most momentous democratic decisions often follow an emotional path rather than a careful weighing of data, such as when the British people in 2016 voted to leave the European Union. Despite warnings from economists, who explained that this decision might ruin the economy, anti-immigrant sentiment and national pride won out. The next day the British pound had its worst drop on record.

Most astonishing are the euphemisms with which we surround the twin driving forces behind human politics: leaders' lust for power and followers' hankering for leadership. Like most primates, we are a hierarchical species, so why do we try to hide this from ourselves? The evidence is all around us, such as the early emergence of pecking orders in children (the opening day at a daycare center may look like a battlefield), our obsession with income and status, the fancy titles we bestow on each other in small organizations, and the infantile devastation of grown men who tumble from the top. Yet the topic remains taboo. Because of my profession, I get to see many social psychology textbooks, and every time a new one arrives, I search the index for the terms *power* and *dominance*. I rarely find them. Apparently, they don't matter. Once when I highlighted the human power drive at a psychology conference, the disapproving comments took me aback. You'd think I had shown them pornography! Attempts to conceal the power motive also emerged in a Dutch study that asked corporate managers about their need to be in control. Whereas they all recognized the hankering for power, none of them applied this insight to themselves. They described their own role in the company in terms of responsibility, prestige, and authority. Invariably, the power grabbers were *other* men.[2]

Political candidates are equally reluctant. They sell themselves as public servants, participating in modern democracy only to fix the economy or improve education. The word *servant* is obvious double-speak. Does anyone truly believe that they join in that mudslinging for our sake? This is why it is so refreshing to work with chimpanzees: they are the honest politicians we all long for. Observing them jockey for posi-

Chimpanzee males are enormously driven to reach the top spot. An alpha male (*left*) looks twice as large as his rival (*right*), even though he is actually the same size. His hair stands on end, and he walks upright in a "bipedal swagger" to impress the other.

tion, one looks in vain for ulterior motives or false promises. It is plain to see what they're after.

The only ones who have addressed our species's longing for power with frankness have been philosophers. Niccolò Machiavelli comes to mind first; Thomas Hobbes postulated an insuppressible power drive, and Friedrich Nietzsche spoke of humanity's "will to power." As a student, realizing that my biology books were of little help explaining chimpanzee behavior, I picked up a copy of Machiavelli's *The Prince*. It offered an insightful, unadorned account of human behavior based on real-life observations of the Borgias, the Medici, and the popes. The book put me in the right frame of mind to write about ape politics at the zoo. To this day, however, people hold their noses while discussing the Florentine philosopher, whom they associate with devious and unscrupulous politics. We are better than that, they seem to be saying, ignoring all evidence to the contrary.

The depth of the human desire for power is never more obvious than in individuals' reactions to its loss. Full-grown men may relapse into displays of uncontrolled rage more often associated with juveniles whose expectations are unmet. When a young primate or child first notices that its every wish will not be granted, a noisy tantrum ensues: this is not how life was supposed to be. Air is expressed with full force through the larynx to wake up the entire neighborhood to this grave injustice. The juvenile rolls around screaming, hitting its own head, unable to stand up, sometimes vomiting, thus imperiling all recent nutritional investment. Tantrums are common around weaning age, which for apes is around four and for humans around two. The reaction of political leaders to the loss of power is very similar, which is why in English we say that they're being "weaned from power." The day Richard Nixon realized he would have to resign the next day, he got down on his knees, sobbed, struck the carpet with his fists, and cried, "What have I done? What has happened?" As Bob Woodward and Carl Bernstein describe in their 1976 book *The Final Days,* Henry Kissinger, Nixon's secretary of state, comforted the dethroned leader as he would a child, literally holding him in his arms, reciting all his many accomplishments over and over until he calmed down.

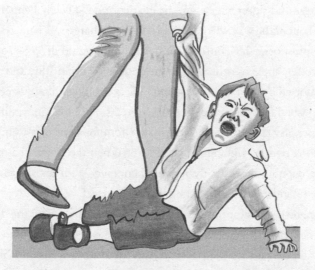

Young primates are expert tantrum throwers when they don't get their way, a phenomenon known to every human parent. Adult primates rarely act like this, except when a male chimpanzee or a human political leader is being weaned from power.

When Steve Ballmer, the head of Microsoft, learned that a senior engineer of his company was leaving to go to work for the competition, he apparently picked up a chair and threw it vigorously across the room. After this outburst, he launched into a tirade about how he was going to f***ing kill those Google boys.³ The higher the emotions run, the more taxing they are for the body. The previous "Dear Leader" of North Korea, Kim Jong-il, is said to have died of a tantrum during an inspection visit to a hydroelectric dam. Since he had ordered repairs, he got extremely upset about a leak. He succumbed to a lethal combination of a bad temper and a weak heart.

For men, as Kissinger once said, power is the ultimate aphrodisiac. They jealously guard it, and if anyone challenges them, they lose all inhibitions. The same occurs in chimps. The first time I saw an established leader lose face, the noise and passion of his reaction astonished me. Normally a dignified character, this alpha male became unrecognizable when confronted by a challenger who slapped his back during a passing charge and slung huge rocks in his direction. The challenger barely stepped out of the way when the alpha countercharged. What to do now? In the midst of such a confrontation, the alpha would drop out of a tree like a rotten apple, writhe on the ground, scream pitifully, and wait to be comforted by the rest of the group. He acted much like a juvenile ape being pushed away from his mother's breast. And like a juvenile, who during a noisy tantrum keeps an eye on Mom for signs of softening, the alpha took note of who approached him. When the group around him was big enough, he instantly regained courage. With his supporters in tow, he rekindled the confrontation with his rival.

Once he lost his top spot, this alpha male sat staring into the distance after every brawl, unaccustomed to losing them. He'd have an empty expression on his face, oblivious to the social activity around him. He refused food for weeks. He became a mere ghost of the impressive big shot he had been. For this beaten and dejected alpha male, it was as if the lights had gone out.

The most cooperative human enterprises, such as large corporations and the military, are those with the best-defined hierarchies. A chain of command beats democracy any time decisive action is needed. We spontaneously switch to a more hierarchical mode when circumstances demand it. In an early study, eleven-year-old boys at a summer camp were divided into two groups to compete against each other. In-group cohesiveness increased, as did reinforcement of social norms and leader-follower behavior. The experiment demonstrated that status hierarchies have a unifying quality that is reinforced as soon as concerted action is called for. This is the paradox of power structures: they bind people together.[4]

Once established, a hierarchical structure eliminates the need for conflict. Obviously, those lower on the scale might prefer to be higher up, but they settle for the next best thing, which is to be left in peace. They also look around for folks who are even lower than themselves to dump their frustrations on. The frequent exchange of status signals reassures bosses that they need not underline their position by force, which gives respite to everyone. Even those who believe that humans are more egalitarian than chimpanzees will have to admit that our societies could not possibly function without an acknowledged order. We crave hierarchical transparency. Imagine the misunderstandings that would arise if no one ever gave the slightest clue about their position in relation to ours. It would be like inviting clerics to a gathering in which an important decision is to be made while asking them to dress in identical garb. Unable to tell priest from pope, the result would be an indecorous commotion, as the higher "primates" would be forced into spectacular intimidation displays—perhaps swinging from the chandeliers—to make up for the lack of color-coding.

Murder

One day in 1980 I received a call that my favorite male chimpanzee, Luit, had been butchered by his own kind at Burgers Zoo. The previous day I had left the zoo worried about him, but now when I rushed back to the zoo, I was totally unprepared for what I found. Normally proud and not particularly affectionate to people, Luit wanted to be touched. He was sitting in a pool of blood, his head leaning against the bars of the night cage. When I gently stroked his head, he let out the deepest sigh. We were bonding at last, but under the saddest circumstances. It was immediately obvious that his condition was life-threatening. He still moved about, but he had lost enormous quantities of blood due to deep puncture holes all over his body. He had also lost some fingers and toes. We soon discovered that he was missing even more vital parts.

As soon as the veterinarian arrived, we tranquilized Luit and took him into surgery, where we sewed literally hundreds of stitches. During this desperate operation, we discovered that his testicles were gone. They had disappeared from the scrotal sac even though the holes in the skin seemed smaller than the testicles themselves, which the keepers had found lying in the straw on the cage floor.

"Squeezed out," the vet concluded dryly.

Luit never came out of the anesthesia. He paid dearly for having stood up to two other males who were frustrated by his sudden ascent. He had stolen their top spots just a few months before, which he was able to do because their coalition had fallen apart. The fight in the night cage marked the sudden resurrection of this coalition, with a fatal outcome.

The night before, the caretakers and I had stayed up late to try to separate three adult males. They all wanted to be in the same night cage, and every time we tried to lower a trapdoor in order to separate them, they either blocked the door with their hands or clung to each other so we couldn't succeed. In the end we gave up and opened all the doors so they would have several interconnected rooms to sleep in.

So the fight that led to Luit's death occurred in isolation from the rest of the colony. We'll never know exactly what transpired. It is not unusual for females to collectively interrupt out-of-control male altercations, but on the night of the assault, the females were in separate night cages within the same building. They must have heard the commotion but been helpless to intervene.

The bloody scene that the caretakers discovered the next morning told us that the other two males had acted together in highly coordinated fashion. They were almost injury free. The younger of the two, who subsequently became the alpha, had a few superficial scratches and bites, but the older male had none at all, suggesting he had held Luit down while letting the younger male inflict all the harm.

Apart from my sadness at losing a male whose character I really liked, and who had been such a wonderful alpha to the colony, the episode of course raised the issue of how much the artificial conditions were to blame. Some commentators said, "What do you want if animals aren't free? Of course they'll kill each other!" As if freedom were the same as being free of stress and strife, which of course it isn't. We know now that the same horrible scenes occur in the wild, but back in 1980 we had no reason to assume a killing like this could happen. The existing sparse reports of lethal aggression in chimpanzees all concerned males of different communities and was typically described as territorial aggression. This is why we had not been overly worried the night before, thinking that these males knew each other well. If they didn't want to be locked up together, they had been given every opportunity to separate.

In retrospect, they desperately stuck together not despite but precisely because of the tensions among them. This sounds counterintuitive, but if power is derived from coalitions, any male who sleeps alone takes a risk. The other two will groom and play and get close, which is exactly what must be avoided. Chimpanzees are very aware of coalitions, both among themselves and among others. They will do anything to prevent hostile coalitions from forming. None of the three males had an interest, therefore, in seeing the other two spend the night without him. And even though Luit had deprived the other two males of the high ranks they once held, he may have been aware that in the end he would need their support rather than their opposition, which meant that he had needed to work on his relationships.

For me, this shocking event had far-reaching consequences. For many nights afterward, I dreamed of the horrible sight that morning. At the same time, I was planning my move to the United States. Somehow these two things got combined in my thoughts, as if Luit had sent me a message about my future. I was busy designing a research agenda for the coming years, weighing the pursuit of all sorts of topics. Would I study aggressive behavior, like almost everyone else, or mate choice, maternal care, intelligence, communication, and so on?

I had just begun to get interested in reconciliation, and instead of looking at this behavior as a luxury that animals can do without, or a "fluke" as some colleagues had called it, I now realized it was absolutely essential. The death of Luit taught me that if the usual methods of conflict resolution fail, things get nasty. I decided to make this the theme of my future research. One way or another, I have dedicated my entire career to it, at first through behavioral observation, then later with experiments on prosocial behavior, cooperation, and fairness. This decision is testimony to the far-reaching influence of emotions. The death of Luit moved me to go with a topic that I thought might provide me with answers but at the time was deemed by many as soft and peripheral.

Only years later did I learn that the incident at the zoo had not been as abnormal as we thought. Even if captive conditions contributed to the way the assault was carried out, they were hardly its cause. The first report of similar behavior in the wild concerned Goblin in Gombe National Park, in Tanzania. Goblin was an exceptional alpha male who acted like a total jerk. In her 1990 book *Through a Window*, Jane Goodall described him as disruptive from a young age, kicking other chimps out of their nests in the early morning for no reason at all. Rather than making friends, he terrorized everyone. One day he got his comeuppance when a mass of angry apes set upon him after he lost a fight against a challenger. The actual attack was hard to see because of the dense undergrowth, but Goblin emerged screaming in panic, with injuries on his wrist, feet, hands, and most important, his scrotum. His wounds were strikingly similar to Luit's.

Goblin would probably have died as his scrotum became infected and swelled, and he developed a fever. Days afterward, he was moving slowly, resting often, and eating little. But a veterinarian darted him and gave him antibiotics. While he convalesced, he stayed out of sight of his own community. Afterward he tried to stage a comeback, directing bluff charges at the new alpha male. This was a grave misjudgment, because it provoked other males in the group to pursue him. Seriously injured again, Goblin was once again saved by the field veterinarian.[5]

Further reports of such violence arrived from the Mahale Mountains, where a Japanese team of primatologists followed the chimps living there for decades. I once visited Mahale, located on Lake Tanganyika, in Tanzania, with my friend Toshisada Nishida, the site's founder, to gain a firsthand impression of chimpanzee politics in the wild. Nishida was a great admirer of the super alpha male Ntologi, who stayed in power for an unprecedented twelve years. Ntologi was a master of divide-and-rule strategies and especially of bribery. For example, even if he hadn't hunted monkeys himself, he'd appropriate the meat of others to distribute it to his supporters while withholding it from his rivals. By controlling the meat flow, he wielded a powerful political instrument. But in the end, this legendary male was aggressively expelled and forced to spend time alone in the periphery of the community territory, limping, hardly able to walk, licking his wounds.

Given the hostility of the neighbors and the helplessness of an isolated male, his situation was perilous. Ntologi did not show his face in the community until he could walk reasonably well again. He would show up and perform a spectacular display of strength and vigor. Then as soon as he was out of sight, he would revert to his limping, wound-licking existence. It was as if he used brief moments of public performance to dispel any notions that his rivals might entertain about his weakened condition, a bit the way the Kremlin in the Soviet Union would parade its ailing leaders on television to make clear that they were not as far gone as rumor had it.

After several comeback attempts and more exiles, Ntologi one day returned as a broken male, forced to accept the lowest of low positions in the hierarchy. Nishida's onetime idol was now a punching bag who ran off screaming whenever a much younger male charged at him. He had lost all his dignity. Then Ntologi was killed in the middle of his community's territory, most likely by a coalition of his own group mates. He was found in a coma covered by deep lacerations, surrounded by chimps who occasionally charged at him. He died the next day.[6]

Several more such cases have been reported, and I would normally stop at this point (I hate dwelling on this kind of behavior), but I cannot omit a most disturbing incident reported a few years ago by the American primatologist Jill Pruetz. Pruetz studies chimpanzees at a rare savanna site in Senegal. The group's alpha male, Foudouko, faced a rebellion after his ally suffered a broken hip. Others in the group took this opportunity to severely attack Foudouko, who was banished to the fringes of their territory. He spent five long years mostly on his own. Every time he tried to return, younger males, perhaps remembering his harsh rule, chased him back out. Then one day Pruetz heard noises about half a mile away from her camp. Approaching the sounds, she faced the troubling spectacle of a dead Foudouko spread out on the ground covered with injuries. The other chimps barely showed any wounds, so their attack must have been highly coordinated. They kept abusing Foudouko's corpse and partially cannibalized him, biting at his throat and genitals and consuming small amounts of flesh. After Pruetz and her co-workers buried Foudouko, the other chimps comforted each other and throughout the night kept making nervous calls in the direction of the grave, as if they were scared of the body.[7]

Ever since the incident at Burgers Zoo, I have looked at the old male, Yeroen, as a murderer. He was the most calculating chimp I have known, a true Machiavellian. He was a great leader so long as he was firmly in the saddle, but became ruthless if anyone stood in his way. I'm sure he was behind the attack on Luit, using the younger male as a pawn. To call an animal a "murderer," however, is of course not something we typically do, because the term implies premeditation.

Many animals kill each other in the heat of battle, such as two stags whose antlers get entangled or a male baboon whose long canines can cause such deep gashes in a rival that blood loss and infections take their toll. In most cases, it is unclear that the other's death was their intention. But when it comes to chimpanzees, the word I hear most often from those who have witnessed assaults on their own kind is how "intentional" their behavior looked. They speak in shocked tones of the extreme brutality, such as attackers who drink the blood of their victims or deliberately try to twist out a leg. Chimps seem determined to terminate the other's life, and they keep going until this goal is achieved. Reportedly they often return to the bloody scene of the "crime" days later, perhaps to verify the effectiveness of their handiwork and make sure of their rival's demise. Finding the body of their victim where they left it, they show no surprise or alarm, which can only mean they expected to see it.

We shouldn't be too surprised by intentional killing as predators do it all the time to different species than their own, which is why we don't call it murder. A predator often doesn't relent until the prey's last breath. When a tiger suffocates a massive gaur—an Indian bison—by clamping its jaws around its throat; when an eagle drags a mountain goat off a cliff so that it falls to its death; or when a crocodile drowns a zebra in the river with a powerful "death roll," they kill their prey deliberately. If the prey shows any sign of life, the predator will resume its assault. Chimpanzees bring the same kind of intentionality to killing their own kind, which is why I don't feel the term *murder* is out of place.

In my experience, the better the leader, the longer his reign will last, and the less likely it will end brutally. We don't have good statistics on this, and I'm aware of exceptions, but generally a male who stays on top by terrorizing everyone else will reign for only a couple of years and end about as badly as Benito Mussolini. With a bully for a leader, the group seems to wait for a challenger and eagerly support him if he stands a chance. In the wild, bully males are expelled or killed, like Goblin and Foudouko, whereas in captivity, they may have to be taken out of the colony for their own safety. Popular leaders, on the other hand, often stay in power for an extraordinarily long time. If a younger male challenges this kind of alpha, the group will side with the latter. For the females, there is nothing better than the stable leadership of an alpha male who protects them and guarantees a harmonious group life. This is the right environment to raise their young in, so females generally want to keep such a male in the saddle.

If a good leader loses his position, he is rarely expelled. He may drop just a few notches on the ladder and then age gracefully within the group. He may also still enjoy quite a bit of influence behind the scenes. I have known one such male, Phineas, for many years. After his alpha position was usurped, he settled in third place and became the darling of the juveniles, romping around with them like a grandpa, as well as a popular grooming partner for all the females. The new alpha permitted Phineas to settle disputes in the colony, not bothering to do so himself because the old male was exceptionally skilled at it. During these years, Phineas was the most relaxed I've ever seen him, which is perhaps understandable because, even though everyone thinks it must be great to be alpha, it is actually a stressful position.

Physiological proof that being the alpha male is not all roses came from baboon droppings collected on the plains of Kenya. Extraction of stress hormones from feces has shown that low-ranking males are much more stressed than high-ranking ones. This sounds logical, because subordinates are chased around and excluded from contact with females. The big surprise, however, is that the top male is just as stressed as the males near the bottom of the hierarchy. This applies only to the highest-ranking male, as he is constantly on the lookout for signs of insubordination and collusion that might unseat him.[8] "Uneasy lies the head that wears a crown," wrote Shakespeare about King Henry IV, which may apply equally to alpha male baboons and chimpanzees.

Drums of War

Even though we share many emotions with other species, we invariably discuss only a subset of them, the "nice" ones, especially if we are proposing these emotions as a reason to value animals more. No one asks us to care about animals because they ferociously assault their enemies or devour their prey. The awareness-raising arguments are always about attachment, mutual assistance, sacrifice, tending offspring, grieving, and the like. In modern times, one of the first books to stress these capacities was Jeffrey Moussaieff Masson's *When Elephants Weep: The Emotional Lives of Animals*, in 1994. We are also reminded of them in the fine works of Elizabeth Marshall Thomas, Temple Grandin, Barbara King, Marc Bekoff, Carl Safina, and others. My own books on primate peacemaking and empathy fit the same mold. But animal emotions undeniably include those that lead them to attack rivals out of sexual jealousy, fight over rank, expand territories at the expense of others, carry out infanticide, and so on. Animal emotional life isn't always pretty.

Our discussions would be more realistic if we considered the entire behavioral spectrum. The first animal emotion studied—the only one that mattered to biologists in the 1960s and '70s—was aggression. In those days, every debate about human evolution boiled down to the aggressive instinct. Without mentioning emotions per se, biologists defined "aggressive behavior" as behavior that harms or intends to harm members of the same species. As always, the focus was on the outcome. But behind aggression was an obvious emotion, known as anger or rage in humans, which also drives animal antagonism. Its bodily manifestations are the same across species, such as low-pitched threatening sounds (grunts, roars, growls). These sounds are associated with body size: the longer the animal's larynx, the deeper the sound. We don't need to see a barking dog to know if it is the size of a Rottweiler or a Chihuahua. Similarly, a male gorilla's chest beat tells us something about the circumference of his torso. During threats, animals inflate their bodies by raising shoulders, arching backs, spreading wings, and puffing up hair or feathers. They show off weaponry, such as claws, antlers, and teeth.

The males of our own species hold up their fists while thrusting out their chest to show their pectorals. The descent of the larynx at puberty in boys but not girls deepens the voice to make men sound big and strong. The purpose of these features is to intimidate and induce fear, so the aggressor will get his way. Most of the time it is effective, but of course if the goal is not reached, things may escalate. Anger is typically aroused by thwarted goals or by challenges to one's status or territory. Showing anger is a common way of getting what one wants and defending what one already has.

Anger and aggression are sometimes described as antisocial emotions, but they are in fact intensely social. If you were to plot on a city map all instances of shouting, insulting, screaming, door-slamming, and china-throwing, they would overwhelmingly be concentrated in family residences: not in the streets, or sport stadiums, or schoolyards, or shopping malls, but inside our homes. Whenever the police try to solve a homicide, their first suspects are family members, lovers, and close colleagues. Since aggression serves to negotiate the terms of social relationships, this is where it typically occurs.

At the same time, close social relationships are also the most resilient. The reason human families manage to stick together is that reconciliation, too, is most common in these relationships. Spouses, siblings, and friends constantly go through cycles of conflict and reconciliation, repeated over and over, to negotiate their relationships. You show anger to make your point, then bury the hatchet with the help of a kiss and some cuddling. Other primates do the same thing to protect their bonds against the eroding effects of conflict: they kiss and groom after fights. For them, too, reconciliation is easiest with the ones they are closest to.

There is one domain, though, in which aggression is common and reconciliation rare, making for decidedly different outcomes. This domain received enormous attention in 1966 when Konrad Lorenz argued in *On Aggression* that we have an aggressive drive that may lead to warfare, hence that war is part of human biology. Many found this view hard to swallow coming from an Austrian who had served in the German army during World War II. The heated and often ideological debate that ensued continues to this date. According to some, it's our destiny to wage war forever, whereas others view war as a cultural phenomenon tied to present conditions.

But modern warfare is undeniably several steps removed from our species's aggressive instincts. It is really not the same thing. The decision to go to war is typically made by older men in a capital based on politics, economics, and egos, while younger men are told to do the dirty work. When I look at a marching army, therefore, I don't necessarily see the aggressive instinct at work. I rather see the herd instinct: thousands of men and women in lockstep, willing to obey orders. I can't imagine that Napoleon's soldiers froze to death in Russia in an angry mood. Nor have I ever heard American Vietnam veterans say that they went over there with rage in their hearts. But alas, the incredibly complex issue of human warfare is still often reduced to that of an aggressive instinct.

In recent history, we have seen so much war-related carnage that it is natural to think warfare is written into our DNA. British prime minister Winston Churchill certainly thought so, writing, "The story of the human race is War. Except for brief and precarious interludes, there has never been peace in the world; and before history began, murderous strife was universal and unending."[9] Unfortunately, or perhaps fortunately, there exists little evidence for Churchill's warmongering state-of-nature. Whereas archaeological signs of individual murder go back several hundred thousand years, similar evidence for warfare (such as graveyards with weapons embedded in a mass of skeletons) is entirely lacking from before 12,000 years ago. We have zero evidence for warfare before the Agricultural Revolution, when survival began to depend on settlements and livestock. Even the walls of Jericho—considered one of the first signs of warfare and famous for having come tumbling down in the Old Testament—may have served mainly as protection against mudflows.

Long before these biblical times, our ancestors lived on a thinly populated planet, with only a couple of million people all told. Studies of mitochondrial DNA suggest that about 70,000 years ago, our lineage was at the edge of extinction, living in scattered small bands. These are hardly the conditions that promote continuous warfare. Nomadic hunter-gatherers—often proposed as models of how our ancestors must have lived—frequently engage in amicable trade, intermarriage, game exchange, and communal feasts. In one of the most striking analyses of recent years, the friendships of the Hadza in Tanzania were mapped, showing that they enjoy a vast network of contacts, well beyond their own group and kin.[10] Even if war was always an option for our ancestors, they probably followed the pattern of the Hadza, which is the opposite of what Churchill surmised. Most likely, long periods of peace and harmony were broken by brief interludes of violent confrontation.

From the beginning, apes have figured prominently in this debate. At first, they were seen as the poster children of our peaceful ancestry, because all they were thought to do was travel from tree to tree in search of food, like a frugivorous version of Rousseau's Noble Savage. In the 1970s, however, came the first shocking field reports of chimpanzees killing each other, hunting monkeys, eating meat, and so on. And even though killing of other species was never the issue, the chimpanzee observations were used to make the point that our ancestors must have been murderous monsters. Incidents of chimps killing their leaders, such as described above, are exceptional compared to what they do to members of other groups for whom they reserve their most brutal violence. As a result, ape behavior moved from serving as an argument against Lorenz's position to becoming Exhibit A in its favor. British primate expert Richard Wrangham concluded in *Demonic Males: Apes and the Origins of Human Violence*, "Chimpanzee-like violence preceded and paved the way for human war, making modern humans the dazed survivors of a continuous, 5-million-year habit of lethal aggression."[11] Wrangham returned us squarely to the idea that warfare is innate, even though he did his best to depict it as a flexible trait, an option subject to our choice. But how flexible can a trait be that has given us "continuous" warfare throughout human history and prehistory?

Although this claim sounds factual, it lacks archaeological backing. We really don't know if war goes back all the way to our earliest ancestors. It is also unclear that these ancestors looked anything like chimpanzees. Due to poor fossilization in the rain forest, the shape and size of our forebears is unknown. That they were apes is an excellent guess, but not only has our own lineage changed since then, so have all the other apes. None of the species around today tell us where we came from. For all we know, the last common ancestor of humans and apes—popularly known as the "missing link"— may have resembled a chimp, a bonobo, a gorilla, an orangutan, or something else. Some experts bet on the gibbon, which

Human   Ardipithecus   Chimpanzee   Bonobo

Of all the great apes, the arm-to-leg ratio of bonobos is most human-like. It is strikingly similar to that of our ancestor *Ardipithecus*, as can be seen in these four hominid silhouettes (not to scale). If we do indeed derive from a bonobo-like ape, human prehistory will have to be rewritten with less emphasis on aggression and more on sex and female power.

is an ape, too, but not one of the "great" knuckle-walking ones. Gibbons are brachiators, who swing through trees hanging by their arms.

Of these possibilities, the bonobo is perhaps the most intriguing given its peaceful predisposition. Although there are lots of confirmed reports of one chimpanzee killing another, thus far there are none for bonobos doing the same, either in captivity or in the wild.[12] On the contrary, fieldworkers describe nonviolent mingling between bonobo communities. The apes call each other when they meet, and may show initial aggression, but soon walk up to each other, have sex, and groom. Mothers let their offspring venture out to play with juveniles of the other group or even with adult males. Bonobos probably have social networks that stretch well beyond their residential communities. Members of different groups seem glad to see each other and act totally relaxed. A recent field report even documented the sharing of meat across territorial borders.[13] This would be unimaginable in chimpanzees, who only know various degrees of hostility. They never show the cordiality and trust that bonobo groups do, and during an intergroup encounter, every chimp mother will try to get as far away as she can because her youngest offspring is at grave risk. This is the stark contrast between our two closest primate relatives: chimpanzee communities in the forest engage in bloody battles, whereas bonobos enjoy happy picnics.

At the Lola ya Bonobo Sanctuary near Kinshasa, in the Democratic Republic of the Congo, it was recently decided to merge two groups of bonobos that had lived separately, just to stimulate some social activity. No one would dare doing such a thing with chimpanzees as the only possible outcome would be a bloodbath. The bonobos produced an orgy instead. Because bonobos freely help strangers to reach a goal, researchers call them *xenophilic* (attracted to strangers), whereas they consider chimpanzees *xenophobic* (fearing or disliking strangers).[14] The bonobo brain reflects these differences. Areas involved in the perception of another's distress, such as the amygdala and anterior insula, are enlarged in the bonobo compared to the chimpanzee. Bonobo brains also contain more developed pathways to control aggressive impulses. The bonobo may well have the most empathic brain of all hominids, including us.[15]

Interesting, you'd think—but science refuses to take bonobos seriously. They are simply too peaceful, too matriarchal, and too gentle to fit the popular storyline of human evolution, which turns on conquest, male dominance, hunting, and warfare. We have a "man the hunter" theory and a "killer ape" theory; we have the idea that intergroup competition made us cooperative, and the proposal that our brains grew so large because women liked smart men. There is no escape: our theories about human evolution always turn around males and what makes them successful. While chimpanzees fit most of these scenarios, no one knows what to do with bonobos. Our hippie cousins are invariably hailed as delightful, then quickly marginalized. *Charming species, but let's stick with the chimpanzee,* is the general tone.

In 2009, when *Ardipithecus ramidus,* a 4.4-million-year-old hominin fossil, was described, her dentition failed to fit the standard narrative. "Ardi's" small canine teeth suggested a relatively pacific species. You'd think this would have been the perfect moment to turn to the bonobo, which is also peaceful and has reduced, blunt teeth. Of all the apes, the bonobo looks most like Ardi, down to its general body proportions, long legs, grasping feet, and even brain size. But instead of offering a new perspective stressing humanity's gentle and empathetic potential along with that of one of its closest relatives, anthropologists gave us only handwringing about how atypical Ardi was—how could we have had such a gentle ancestor? Presenting Ardi as an anomaly and a mystery kept intact the prevailing macho storyline.

Accordingly, humans in a "state of nature" (if such a condition ever existed) wage continuous war. Our only hope is civilization, as Steven Pinker writes in *The Better Angels of Our Nature: Why Violence Has Declined,* a 2011 book that favors chimpanzees as the best model to understand where we come from. Pinker offers cultural progress as the solution to all our problems. We need to bring our instincts under control—otherwise we'd act like chimps. This distinctly Freudian message (Sigmund Freud saw civilization as the tamer of our basic instincts) is deeply ingrained in the West and remains immensely popular. In the meantime, however, cultural anthropologists and human rights organizations abhor the inevitable implication that preliterate people live in chronic violence. This myth can be (and has been) used as an argument against these peoples' rights. Perhaps a handful of tribes behave this way, but critics have argued that only serious cherry-picking from the anthropological record can support Pinker's blood-soaked view of human origins. "Savages" are not nearly as savage as is often assumed.[16]

The most puzzling part of the whole civilization-to-the-rescue proposal is that whenever modern-day explorers have encountered preliterate people, the violent ones have invariably been the former. This was true when the British discovered Australia, when the Pilgrims landed in New England, and when Christopher Columbus came to the New World. Even if the indigenous people greeted the foreign visitors with gifts and friendship, all the latter could do was massacre their hosts. Columbus encountered people who didn't even know what a sword was, only to marvel that with just fifty soldiers he'd be able to crush them. So much for civilization's edifying influence.[17]

My own views focus not on human history but on the natural abilities of primates to dampen conflict. Most of the time they are excellent at keeping the peace. I can't believe that we are still bowing to Freud and Lorenz, not to mention Hobbes, while debating our evolutionary background. The idea that we can achieve optimal sociality only by subduing human biology is antiquated. It doesn't fit with what we know about hunter-gatherers, other primates, or modern neuroscience. It also promotes a sequential view—first we had human biology, then we got civilization—whereas in reality the two have always gone hand in hand.

Civilization is not some outside force: it is us. No abiological humans ever existed, nor any acultural ones. And why do we always consider our biology in the bleakest possible light? Have we turned nature into the bad guy so that we can look at ourselves as the good guy? Social life is very much part of our primate background, as are cooperation, bonding, and empathy. This is because group living is our main survival strategy. Primates are made to be social, made to care about each other, made to get along, and the same applies to us. Civilization may do all sorts of great things for us, but it does so by co-opting natural abilities, not by inventing anything new. It works with what we have to offer, including an age-old capacity for peaceful coexistence.

Ardi is telling us something, and even if we don't agree about what she's saying, it's time for us to start discussing human evolution without always playing the drums of war in the background. True, on our bad days, we are as domineering and violent as chimps can be, but on our good days, we are as nice and sensitive as bonobos can be.

Female Power

Mama was the top-ranking female in the Arnhem chimpanzee colony. Even though she didn't physically dominate any grown males, she was more powerful and influential than most. I have known other impressive female chimps who knew how to stand their ground, pushed males around (taking food out of their hands, nudging them out of a comfortable seat), and were so pivotal to group life that everyone courted their political backing.

But the true champions in this regard are not chimpanzees but bonobos. In the wild, an alpha female bonobo will stride into a clearing dragging a huge branch behind her, making a display that everyone else watches but avoids. It's not unusual for female bonobos to chase off males and lay claim to large fruits that they divvy up among themselves. *Anonidium* fruits weigh up to twenty-two pounds, and *Treculia* fruits up to sixty-six pounds, nearly the weight of a grown bonobo. Once these colossal fruits drop to the ground, females invariably claim them and rarely see fit to share with begging males. Individual males may supplant individual females, especially younger ones, but collectively the females dominate the males.[18]

This is true not only in the wild but also in every zoo that I have visited. Invariably, a female is in charge of the bonobo colony. The only exception occurs if a zoo has just one male and one female. Male bonobos are bigger and stronger and possess longer canine teeth than females. In these cases, the male is boss. But as soon as the colony grows, and the zoo adds a second female, male supremacy is over. The females will band together whenever he tries to intimidate one of them. Adding more males won't change much, because unlike chimpanzee males, which easily form coalitions, bonobo males aren't very cooperative.

Naturally, bonobos have become popular among feminists, who have honored them in book dedications, such as Alice Walker thanking Life for their existence. Whereas chimpanzees resolve sexual issues with power, bonobos resolve power issues with sex. Moreover, they have sex in all possible combinations, including members of the same sex. Scientists and journalists have pushed back, though, feeling that the bonobo is just too good to be true. Could this make-love-not-war species be a "politically correct" concoction, a made-up ape to satisfy the liberal left? One journalist traveled all the way to the Democratic Republic of the Congo in an attempt to prove that bonobos aren't nearly as pacific as they have been made out to be. All he came back with, however, was the story of a bonobo chasing a duiker. The little antelope escaped safely, but the journalist felt the need to treat his readers to a ghastly account of the way the ape *might* have killed and devoured it. None of this had much to do with the issue at hand, however, because predation is not aggression. Predation is driven by hunger, not competition.

Scientists unfamiliar with the species have criticized those who dare call male bonobos subordinate. Better to regard them as "chivalrous," they said, because the weaker sex's clout obviously depends on the stronger sex's good-heartedness. Furthermore, female dominance cannot be all that important since it is limited to food. This was a baffling twist, because if there's one yardstick that has been applied to every animal on the planet, it is that if individual A can chase B away from its food, A is called dominant. The bonobo pioneer Takayoshi Kano, the Japanese fieldworker who followed bonobos for twenty-five years in the swamp forest, noted that food is precisely what female dominance is all about. If this is what matters to them, then it should matter to the human observer, too. Kano went on to point out that even if there is no food around, full-grown males react submissively to the mere approach of a top-ranking female.[19]

Collective female rule is especially surprising given that bonobo females are unrelated to each other. They are friends, not kin. At puberty, a female leaves the community in which she was born to join the neighbors, where she attaches herself to a senior female, who takes her under her wing. She generally lacks relatives in the territory in which she takes up residence. I have called the resulting gynarchy a *secondary sisterhood*. The females act with sisterly solidarity, based on common interest rather than kinship. In recent years, we have learned more about these networks thanks to a revival of research on wild bonobos, which for many years was disrupted by war. Field observations are extremely hard to conduct in one of the remotest forests in the world, but the Japanese scientist Nahoko Tokuyama managed to collect crucial information on how the females work together. Most of the time, they do so in response to male harassment. Whereas chimpanzee females endure abuse and the occasional killing of their infants, bonobo females suffer none of these problems. High-ranking older females support younger ones as soon as they are in trouble with any males. Female bonobos have a relatively carefree existence thanks to this camaraderie, which keeps male violence in check.[20]

We know much less about the dominance relationships among the females themselves. They usually have a clear alpha, who in the bonobos' case also is the general alpha individual, dominant over everyone. But competition over this position is less intense than that among males in chimpanzee society. That's because for females there is always less at stake than for males. In evolutionary terms, all that matters is who passes on their genes. Males can do better in this regard than any female, because a top position allows them to impregnate a multitude of females. For females, the evolutionary game is radically different. Regardless of rank and number of mates, a female still gets only one infant at a time. Because of the way reproduction works, male status carries a bigger premium.

Bonobo females, however, do the next best thing. They are fierce supporters of their sons in the male hierarchy. The worst fights in bonobo society occur when females get involved in their sons' status struggles. Male bonobos vie for positions on their mothers' apron strings, which allows high-ranking mothers to increase their number of grandchildren. Take the wild alpha female Kame, with no fewer than three grown sons, the oldest of whom was the alpha male. When old age began to weaken Kame, she became hesitant to defend her children. The son of the beta female must have sensed this because he began to challenge Kame's sons. His own mother backed him up and was not afraid to attack the top male on his behalf. The frictions escalated to the point that the two mothers exchanged blows, rolling over the ground. The beta female held Kame down. Kame never recovered from this humiliation, and soon her sons dropped to midranking positions. After Kame's death, they became peripheral, and the son of the new alpha female took the top spot.[21]

The closest human parallel is the fierce competition and intrigues among slave concubines in the Ottoman Imperial Harem, some of whom gained a status equal to the sultan's wives. These women groomed their sons to become the next sultan. Upon ascension to the throne, the winner would inevitably order the killing of all his brothers, so he would be the only one to sire offspring. We humans simply do things more radically than bonobos.

Even though power will always be a greater obsession for males than for females, the will to power is not restricted to males. Yet it cannot be denied that in our own societies women with political ambitions face special challenges. One is that while being attractive and good-looking is great for men (think of John F. Kennedy or Justin Trudeau), it doesn't work out equally well for women. This is related to how sexual competition interacts with an electorate that is half male and half female. Attractive women, especially those of childbearing age, are perceived as rivals by other women, which makes it hard for them to get their vote. When John McCain ran against Barack Obama in 2008, he selected a relatively young woman, Sarah Palin, as his running mate. Men in the media regarded it as a brilliant move, calling Palin "hot" and a "MILF," but no one seemed to realize how much male enthusiasm might harm Palin's standing among women. Obama barely won the male vote (49 to 48 percent), but he ran away with the female vote (56 to 43 percent).

Women begin to appeal as leaders only after they have become invisible to the male gaze by leaving their reproductive years behind. Modern female heads of state have all been postmenopausal, such as Golda Meir, Indira Gandhi, and Margaret Thatcher. The most powerful woman of our era, Angela Merkel of Germany, doesn't even like to draw attention to her gender, dressing as neutrally as possible. Merkel is a skilled and shrewd politician who is unimpressed by men. When Vladimir Putin received her at his Russian dacha in 2007, he introduced his large pet Labrador to her, knowing full well that Merkel was scared of dogs. In the end, his tactic failed, because she drew a distinction between Putin and his dog, noting to journalists, "I understand why he has to do this—to prove he's a man. He's afraid of his own weakness."[22] Putin's tactic showed how men always seek the upper hand through intimidation.

That removing a male from power triggers the same reaction as yanking the security blanket away from a baby proves how deeply ingrained these tendencies are. We tend to underestimate the emotions that organize our lives and institutions, but they are at the core of everything we do and are. The desire to control others is a driving force behind many social processes and imposes structure on primate societies. From Trump's and Clinton's quest to lead the nation to bonobo mothers who come to blows over their sons, the power motive is pervasive and plain to see. It has led to some of our highest achievements under inspiring leaders, but it also has a disturbing track record of violence, including the political assassinations to which our own species is no stranger.

Emotions can be good, bad, and ugly, which is as true for animals as it is for us.

6 | EMOTIONAL INTELLIGENCE

On Fairness and Free Will

In a picture of a serene savanna, a zebra stands with his rump turned toward the viewer. He lifts his head to get a better look at two out-of-focus lions in the distance, which are in the midst of intercourse. My Facebook friends suggested captions, the best one having the lioness exclaim: "Wish you'd hurry up, Arthur! I've just spotted our supper!"

But copulating lions are not in the mood for supper. The zebra knows this, which is why he is in no hurry to trot off. He has no fear, at least not at this moment. Fear is a self-protective emotion, which puts it at the top of the survival value list. But even fear is aroused only after careful judgment of a situation. Just seeing lions doesn't do it. Antelopes, zebras, and wildebeest act quite relaxed around big cats who are lying around, at play, or having sex. They are familiar with cat behavior and know perfectly well when their enemies are in a hunting mood. That's when they get afraid, if they notice it in time.

Celebrating the Cerebral

Emotion-based reactions have this gigantic advantage over reflex-like behavior: they pass through a filter of experience and learning known as *appraisal*. I wish early ethologists had thought of this, instead of clinging to the instinct concept, which is now largely outdated. Instincts are knee-jerk reactions, which are pretty useless in an ever-changing world. Emotions are much more adaptable, because they operate like intelligent instincts. They still produce the desired behavioral change, but only after a careful evaluation of the situation. This evaluation may take only a fraction of a second, yet depends on comparing current conditions with past experience, such as the zebra on the savanna is doing. If I'm planning a picnic, for example, the sight of rain will depress me, but if I intend to stay home, my wet Dutch background kicks in: I love seeing rain outside my window. Banging noises from a car muffler freak out those who have lived through military combat, while others barely notice. The sound of a barking dog scares us until we see he's on a leash. Emotions always run through an appraisal filter, which explains why different people react differently to the same situation.

We may not be in full control of our emotions, but we aren't their slaves either. This is why you should never say "my emotions took over" as an excuse for something stupid you did, because you *let* your emotions take over. Getting emotional has a voluntary side. You let yourself fall in love with the wrong person, you let yourself hate certain others, you allowed greed to cloud your judgment or imagination to feed your jealousy. Emotions are never "just" emotions, and they are never fully automated. Perhaps the greatest misunderstanding about emotions is that they are the opposite of cognition. We have translated the dualism between body and mind into one between emotion and intelligence, but the two actually go together and cannot operate without each other.

The Portuguese-American neuroscientist Antonio Damasio reported on a patient, Elliott, with ventromedial frontal lobe damage. While

Elliott was articulate and intellectually sound, witty even, he had become emotionally flat, showing no hint of affect in many hours of conversation. Elliott was never sad, impatient, angry, or frustrated. This lack of emotion seemed to paralyze his decision making. It might take him all afternoon to make up his mind about where and what to eat, or half an hour to decide on an appointment or the color of his pen. Damasio and his team tested Elliott in all sorts of ways. Even though his reasoning capacities seemed perfectly fine, he had trouble sticking with a task and especially reaching a conclusion. As Damasio summarized: "The defect appeared to set in at the late stages of reasoning, close to or at the point at which choice making or response selection must occur." Elliott himself, after a session in which he had carefully reviewed all options, said "And after all this, I still wouldn't know what to do!"[1]

As a result of Damasio's insights and other studies since, modern neuroscience has ditched the whole idea of emotions and rationality as opposing forces, like oil and water, that don't mix. Emotions are an essential part of our intellect. The idea that they are separate remains so ingrained, however, that it survives in full force in many circles. People still look down on the emotions and think that sound decision making requires that we be clear-headed and dispassionate, as in Darwin's sham decision process, listing the pros and cons of marriage. This illusion goes back to the ancient Greek philosophers, who greatly admired the logical reasoning of men like themselves but recognized less of this capacity in women and even less in animals. Women were considered sentimental and intuitive, more in tune with their bodies, hence not nearly as intellectual as men. Men were immune to monthly mood swings and moreover were the only ones capable of putting the screws on their passions.

The one thing that bothered the philosophers, though, and that seems to discomfit many of them still, is that the human mind requires a material vessel. It can't exist without a body. The earthbound state of the mind is most unfortunate, because not only does the body pester us with uncontrollable urges and feelings, forcing us to think of things we don't want to think about, it is mortal to boot. As the Gospel of Thomas complains about the human spirit: "I am amazed at how this great wealth has made its home in this poverty."[2]

Disdain for the body explains why medieval hermits—overwhelmingly male—tried to negate it. They'd withdraw to the desert or a nearby cave to deprive themselves of all temptations of the flesh—only to be tormented by visions of lavish meals and voluptuous women. It also explains why rich people—again, overwhelmingly male—line up to have their heads cryogenically frozen after death. Their brains will be ready for the day when technology reaches the point that they can be "uploaded." Believing that the mind doesn't need a body, they pay a fortune for a digitally immortal future in which everything that is currently in their head can be moved to a machine. After all, the mind is just some software that runs on a flesh platform. It could run equally well on a computer. Never mind that science hardly knows what a mind without a body would look like. The computer metaphor for the brain is profoundly misleading given that the brain is wired in a million ways to the body and is an integrated part of it. The human mind makes no distinction between body and brain and represents both. I am not at all convinced therefore that waking up in a digital format will be a happy moment. Happiness is visceral, and a brain severed from the viscera probably doesn't feel anything.[3]

This is what we are up against in any discussion about animal emotions. We have this delusion of a free-floating human spirit that is barely connected to biology, more prominent in one gender, and a radical departure from anything that came before. We celebrate the cerebral, believe in such a thing as "pure reason," and have a low opinion of the emotions, the body, and any species not our own. These cultural and religious prejudices have been with us for millennia, hence are not easily erased. Yet we will need to distance ourselves from them before seriously considering that animal emotions are a sign of intelligence, as I will argue here.

Like humans, animals have *emotional intelligence,* a pop-psychological concept that refers to the ability to read another's emotions, use emotional information, and control one's own emotions in order to achieve goals.[4] I use the term loosely here to denote the interplay between emotion and cognition. We generally study human emotional intelligence as an individual trait. Some of us are better than others at managing emotional upheavals or taking advantage of how we feel. It's all about one's upbringing, skills, and personality. When it comes to animals, though, we consider how emotions and cognition go hand in hand to produce the outcomes we see, from social hierarchies to family life, and from antipredator defense to conflict resolution.

A good illustration is the sense of fairness. It is often considered a product of reason and logic, and a uniquely human moral value, but it would never have arisen without a basic emotion that we share with other primates, canines, and birds. Our sense of fairness is an intellectual transformation of this shared emotion.

Cucumber and Grape Monkeys

For more than twenty years, I maintained a capuchin colony at the Yerkes National Primate Research Center. About thirty of these handsome brown monkeys lived together in an outdoor area connected to our lab, where we tested them every day on social intelligence. We set up situations in which they could work together, share food, exchange tokens, recognize faces, and so on. The monkeys did all this with great enthusiasm. Capuchins are in their best mood when they are busy. They never give up—in the wild, they pound an oyster until the mollusk relaxes its muscle so they can pry it open, and in the lab, they keep selecting the faces of group mates and strangers on a touchscreen until they can tell them apart. They bring persistence to every task. We never forced them to come in, but instead tried to make the sessions short and sweet (literally) so that they'd be eager.

What I perhaps liked most about them was the background noise they produced. While at work, they couldn't stop "talking" to the rest of their group. In the wild, capuchins are forest dwellers, which means that they are often out of each other's sight due to the leaves and trees. Vocalizations are their lifeline to the group. In the lab, we tested individuals away from the rest, but always within earshot. They continuously called to their family and friends and got calls back.

I grew so fond of this species, and the personalities that I knew by name, that I went to see capuchins in Brazil and Costa Rica to get a sense of how they behave in nature. And even though they're just monkeys, almost all their mental capacities are most recognizable to us. I say "just" monkeys facetiously, because experts on ape behavior sometimes talk in this condescending tone about other primates, a bit like how paleontologists can't stand the idea that a newly discovered fossil might be "just" an ape and invariably try to squeeze it into the artificial construct that is the genus Homo.

Capuchins are remarkable monkeys, though, which became clear when it was discovered that they use stones to crack open nuts in the forest. They bring hammer stones and nuts from far away to rock outcrops that serve as anvils. Stone tool technology had been hailed as an extraordinary hominid achievement that we shared only with chimpanzees. Now the lithic club had to admit these little monkeys with prehensile tails. Relative to their cat-size bodies, capuchin brains are as large as those of chimpanzees, however, and they live extraordinarily long lives. Mango, the oldest female in my colony, is still alive today, estimated at fifty.

My former student Sarah Brosnan and I made an accidental discovery in capuchin behavior that shook up traditional notions about human fairness, which was widely considered a cultural rather than a biological phenomenon. We have trouble imagining fairness as an evolved trait partly due to how we depict nature. Using evocative phrases such as "survival of the fittest" and "nature red in tooth and claw," we stress nature's cruelty, leaving no room for fairness, only the right of the strongest. In the meantime, we forget that animals often depend on each other and survive through cooperation. In fact, they struggle far more against their environment or against hunger and disease than against each other. This is why the naturalist and anarchist Pyotr Kropotkin asked in 1902: "Who are the fittest: those who are continually at war with each other, or those who support one another?"[5] Having watched horses and musk oxen in Siberia huddle together against freezing snowstorms or form a protective ring around their young to keep the wolves at bay, the Russian prince opted for mutual aid as the superior strategy. He was way ahead of his time.[6]

Back in the lab, Sarah and I were baffled that our capuchins, instead of just consuming their own rewards, also kept an eye on those of others. This hadn't been noticed before because of how animals are typically tested. A rat sits alone in a box pressing a lever for a reward. The only thing that matters to the rat is how tough the task is, how attractive its reward, and at what schedule the rewards are delivered. Thanks to my interest in social behavior, however, my lab was different. Monkeys were rarely alone during tests. That was how we noticed that they closely eyed every morsel of food that went to others. It was as if they valued their own reward relative to what others got. This may seem ridiculous, because shouldn't they just focus on their own performance and their own payoff?

But in light of what we know about human behavior, their interest in others made perfect sense. The so-called Easterlin Paradox was named after Richard Easterlin, an American economist who noticed that within every society, rich people tend to be happier than poor ones. So far so good, but Easterlin also found that if an entire society grows richer, its average well-being doesn't go up. In other words, people in a wealthy nation do not feel any better than those in a poor one. How is this possible if wealth makes us happy? The answer is that it is not wealth per se that enhances well-being but *relative* wealth. Feelings of happiness depend on how our income stacks up against that of others.[7]

At the time, however, Sarah and I were unfamiliar with the Easterlin Paradox. Having noticed that our monkeys got upset whenever their rewards fell short, we decided to take a closer look. This led to a relatively simple experiment that exploited the talent of capuchins for barter, which they do spontaneously. If you ever forget a screwdriver in their cage, you need only point at the tool and hold up a peanut, and they will hand it to you through the mesh. They are so fond of barter that they will even bring you a dried orange peel in exchange for a pebble, both useless items. More remarkable still, after they place the object in your palm, they may grab your fingers with their tiny hands to bend them inward, thus closing your hand around the object, as if saying, *There you go, hold it tight!*

We used this natural talent, which probably relates to food sharing, to put our monkeys to work. We had noticed that inequity reactions occur only if effort is involved. If you just feed two monkeys different foods, they barely notice, but if both of them work for it, all of a sudden it matters what one gets relative to another. Food has to serve as a salary, so to speak, for inequity to become an issue.

For our experiment, we placed two monkeys in a test chamber sitting side by side with mesh between them.[8] We'd drop a small rock in the area of one of them, then hold up an open hand to ask for the rock back. We'd do this with both monkeys in alternation twenty-five times in a row. If both of them got cucumber slices in return for the rocks, they'd make the exchanges all the time, contentedly eating their food. But if we gave one monkey grapes for the exchanges while keeping the other one on cucumber, we'd trigger some real drama. Food preferences generally match prices in the supermarket, so grapes are far superior to cucumber. Upon noticing their partner's raise, the monkeys who'd been perfectly happy to work for cucumber all of a sudden went on strike. Not only would they perform reluctantly, they'd grow agitated, hurling the pebbles out of the test chamber and sometimes even the cucumber slices. A food that they normally never refused had become less than desirable: it had become distasteful![9]

Exploring the sense of fairness, we conducted a cucumber and grape experiment with brown capuchin monkeys. Two monkeys were placed side by side in a test chamber behind Plexiglas with holes. If both monkeys received cucumber, they performed a simple task almost 100 percent of the time. Inequity was created by giving one monkey grapes, which are much preferred, while the other still received cucumber. Even though her own reward was unchanged, this greatly upset the cucumber monkey. She refused to perform and rejected the cucumber slices, hurling them out of the test chamber.

Their frustration was so intense that we decided to give both monkeys lots of goodies before returning them to their group, to keep them from developing negative associations with the experiment. We obviously didn't use just two monkeys but tested many of them in different combinations before reaching a conclusion.

Throwing perfectly fine food away is what economists call "irrational" behavior. Every rational actor supposedly takes what he or she can get in order to maximize profit. If I give you $1 and your friend $1,000, you may be pissed off, but you still should take your $1, because it's better than nothing. People are not rational maximizers, though. This conclusion is nowadays hailed by dramatically declaring the demise of *Homo economicus*, the image of our species offered by economics textbooks, according to which we make perfectly rational decisions in order to satisfy our greed. Studies have undermined this popular assumption by showing that emotional biases often make us choose quite differently.

We're not nearly as rational and selfish as we're thought to be, and not all our desires are material.

That the same applies to other species, however, is not always recognized. The American anthropologist Joseph Henrich jauntily concluded from his long quest to find *Homo economicus*, "Eventually, we did find him. He turned out to be a chimpanzee."[10] Very funny, but his remark was based on studies from about fifteen years ago in which chimpanzees had failed to show concern for the well-being of others. Even though there are good reasons to doubt negative evidence (following the mantra "Absence of evidence is not evidence of absence"), these early findings had got quite a bit of traction. The ape studies on which they were based, however, have now been surpassed by many others that convincingly demonstrate empathy and prosocial tendencies. In fact, the default inclination of most primates is cooperative, not selfish, so we can safely conclude that *Homo economicus* never evolved, neither in our direct lineage nor anywhere else in the primate order. He is deader than dead.

At first, Sarah and I dodged any talk of "fairness" in monkeys and prudently spoke of "inequity aversion." But when word of our study got out—which happened on the exact same day, in 2003, that the head of the New York Stock Exchange, Richard Grasso, was forced to resign due to a sky-high compensation package that offended the nation—the media recognized the evolutionary roots of fairness, and said fairness must be a good thing if even monkeys have it.

This news upset quite a few people, and we received angry emails complaining that we must be Marxists for trying to demonstrate that fairness is natural, or that it is impossible for monkeys to have the same sense as we do, because fairness was invented during the French Revolution. To me, these were crazy complaints, because I view our monkeys as minicapitalists (they work for food and compare income), and I don't believe in moral principles invented by a bunch of old guys in Paris. Morality goes so much deeper than that.

The sense of fairness is a great example of how we move from "moral sentiments," as the Scottish philosopher David Hume called them, to full-blown moral principles. The starting point is always an emotion. In this case, the emotion is envy. The monkeys begrudged their companion's advantage. It was not simply the sight of high-quality food that upset them, because we also conducted tests in which the grapes did not go to a partner but were merely visible in a bowl or thrown into an empty cage next to them. They responded far less to these situations, which meant that their rejection of cucumber concerned a *social* comparison. It was specifically caused by seeing someone else get something better.

You might say that this still doesn't amount to a sense of fairness, because only one monkey got upset—the lucky one didn't seem to care. This is true, but the first monkey's envious reaction is nevertheless at the core of the sense of fairness, as I will explain. Let me add that similar reactions occur in other species. They are also typical of young children when one of them gets a smaller pizza slice than his sibling (yelling "That's not fair!"). Many a dog owner has approached me to describe

The Clever Dog Lab in Vienna tested the sensitivity of dogs to inequity by asking two of them to give a paw to a human experimenter. Without receiving a reward, both dogs would do so many times in a row. But if one of them received a piece of bread for the action, whereas the other received nothing, the latter (*left*) would give up and refuse.

how one of their pets reacts if the other receives better food. In fact, when dogs were tested at the Clever Dog Lab at the University of Vienna, it was found that they were willing to lift their paw many times in a row for a shake even if they got no reward for it. But as soon as another dog received food for the same trick, the first dog lost interest and refused any further paw shakes. Wolves raised in human homes behaved the same way.[11]

Being resentful about another's success may seem petty, but in the long run it keeps one from getting duped. To call this response "irrational" misses the mark. If you and I often go out hunting together and you always claim the best chunks of meat, I need to either vociferously object to the way you are treating me or else start looking for a new hunting buddy. I'm sure I can do better than that. Sensitivity to reward distribution helps ensure payoffs for both parties, which is essential for continued cooperation. It is probably no accident that the animals most sensitive to inequity—chimps, capuchins, and canids—hunt in groups and share meat.

The resentment response may not be limited to these animals, though. The American psychologist Irene Pepperberg described her typical dinner conversation with two squabbling African grey parrots, the late psittacine genius Alex and his junior colleague Griffin:

> *I had dinner, with Alex and Griffin as company. Dining company, really, because they insisted on sharing my food. They loved green beans and broccoli. My job was to make sure it was equal shares, otherwise there would be loud complaints. "Green bean," Alex would yell if he thought Griffin had had one too many. Same with Griffin.*[12]

But still, none of this goes beyond the egocentric reaction that we have dubbed *first-order fairness,* marked by irritation at being shortchanged compared to somebody else. Only after we began working with apes did we find signs of *second-order fairness*, which concerns equity in general. Humans don't balk just at getting less than another but sometimes also at getting more. We may feel uncomfortable about our own advantage. Not that Grasso showed much of this sensitivity—it's rather weakly developed in our species—but in principle fair outcomes are sought not only by the poor but also by the rich.

Second-order fairness can be seen in the natural behavior of apes, such as when they resolve conflicts over food that is not theirs. I once saw two juveniles quarreling over a leafy branch. An adolescent female chimp interrupted the quarrel, took the branch from them, broke it in two, then handed each a part. Did she just want to stop the fight, or did she understand something about distribution? High-ranking males, too, often break up fights over food without claiming any of it for themselves. They just settle the dispute, which allows all parties to share. Panbanisha, a bonobo, being tested in a cognition laboratory, earned large amounts of milk and raisins in exchange for a task she performed. But her friends and family were following everything from a distance, and she felt their envious eyes on her. After a while, Panbanisha began to refuse rewards, as if worried about being privileged. Looking at the experimenter, she kept gesturing to the others until they, too, got some of the goodies. Only when they got some did she eat hers.[13]

Apes can think ahead. Had Panbanisha publicly eaten her fill, there might have been unpleasant consequences when she rejoined the others later on.

The Ultimatum Game

Affluent people are known to quietly remove the price tags from the furniture, kitchen tools, and other expensive items that they bring home so as not to upset their nannies and household staff. They are reticent to show off. When the sociologist Rachel Sherman interviewed wealthy New Yorkers, she found that they were uncomfortable with income disparities and made attempts to soften them. They'd avoid calling themselves "rich" or "upper class" and preferred being considered "fortunate."

They seemed to realize that their situation might arouse resentment, something they'd like to avoid.[14]

This is a good start, but the removal of price tags is a mere Band-Aid—it doesn't fool anyone. The only efficient way to avoid envy is the one Panbanisha chose, which is to share one's wealth. This is common in small-scale human societies—hunter-gatherers have an extreme sharing ethos, not even allowing successful hunters to brag about their skills. The same applies to chimpanzees. This first dawned on us during a large-scale study by Sarah Brosnan in which she rewarded apes with a carrot piece for performing a simple task. Occasionally, however, she would give one of them a grape, the favorite reward.

As with the monkeys, carrot recipients, if their partner received a grape, refused to perform the task or threw out their reward. But no one had anticipated that the grape recipients, too, might be troubled. They sometimes refused the grape if their partner received only a carrot, but not if their partner also got a grape. Since this comes much closer to the human sense of fairness, we developed a bold plan to play the Ultimatum Game with chimps. Considered the gold standard in assessing human fairness, this game has been played with people all over the world.

At the outset, one person is given, say, one hundred dollars to split with another. This person decides the proportion of the split: it may be fifty-fifty, or it may be any other division, such as ninety-ten. The partner may accept the offer, in which case both players get money. But the partner may also refuse the offer, leaving both empty-handed. The partner's veto option means that the person who divides the dough better watch out, because the partner may not like a lowball offer.

Obviously, if humans were rational maximizers, the partner would never refuse the offer but would accept *any* deal. But even people who have never heard of the French Revolution reject excessively low offers. The more a culture relies on cooperation, the more likely its members are to reject low offers. Whale hunters from Lamalera, Indonesia, for example, roam the open ocean in large canoes holding a dozen men each. They capture whales by jumping onto the leviathan's back and thrusting a harpoon into it. Entire families depend on the success of this extremely hazardous activity, so when the hunters bring home a whale, distribution of the bonanza is very much on their mind. Not surprisingly, these hunters are more sensitive to fairness than most other cultures, such as horticultural ones in which every family tends its own plot of land. The human sense of fairness is closely tied to cooperation.[15]

Chimpanzees cooperate too, both in the hunt and in territorial defense, not to mention political alliances. But how can we play the Ultimatum Game with other species when we can't explain to them how it works? Our solution was to use tokens painted in either of two colors that they could exchange for food in corresponding amounts. Our co-worker, Darby Proctor, invited two individuals to sit next to each other separated by bars, and offered one of them a choice between two different-colored tokens. If the chooser picked one color, Darby would give her five slices of banana and her partner only one. If the chooser chose the other color, Darby would give each of them three slices. So the chooser faced a simple decision between an outcome that was better for herself and one that was best for both. The important part was that, as in the Ultimatum Game, the chooser's partner had to "agree" to the choice. The chooser could not return the token directly to Darby—it had to come from the partner. So the chooser had to pass the token through the bars to her partner, who had to accept it by returning it to Darby.

The chimps quickly learned the meaning of the two colors, as was clear from the way the partner reacted when a chooser offered them the selfish token, the one giving the chooser five times more. The partner would bang on the bars between them or spit mouthfuls of water at the other to express displeasure.

When Darby played the same game with preschool children—rewarding them with stickers rather than banana slices—they showed similar reactions, but verbalized. Getting the less attractive token, they'd say, "You got more than me" or "I want more stickers!" Except for this difference in expression, apes and children behaved the same way. In the majority of trials, choosers preferred the token that produced equal rewards. At first sight, this decision seems costly, but not if we factor in the value of social relationships. Being too selfish might cost a friendship.[16]

If you now ask me if there is any difference between the human sense of fairness and that of chimpanzees, I really don't know anymore. There are probably a few differences left, but by and large both species actively seek to equalize outcomes. The great step up compared with the first-order fairness of monkeys, dogs, crows, parrots, and a few other species is that we hominids are better at predicting the future. Humans and apes realize that keeping everything for themselves will create bad feelings. So second-order fairness can be explained from a purely utilitarian perspective. We are fair not because we love each other or are so nice but because we need to keep cooperation flowing. It's our way of retaining everyone on the team.

This is what I mean by emotional intelligence. The human and ape sense of fairness starts with a negative emotion but then combines it with a realization of its harmful effects and turns it into something positive. "Thou shalt not covet" is great advice, but even better is to remove the reasons for coveting. My view here is quite the opposite of the one espoused by the American moral philosopher John Rawls in *A Theory of Justice*, his celebrated 1971 treatise on this topic. While I admire Rawls's sophisticated arguments for why justice is better than injustice, his philosophizing neglects our species's emotional core. He considers only emotions he approves of, declaring toward the end of his book that "for reasons of both simplicity and moral theory, I have assumed an absence of envy."[17]

I am stupefied. Since when can we simply drop an emotion from an analysis of human behavior? Who in his right mind would even do such a thing, especially for an emotion so ubiquitous that it is known in every language? Envy even has a color—the "green-eyed monster,"

Shakespeare called it in *Othello*. Rawls believes that principles of justice should be deliberately chosen by people free of envy. But where do we find such people? And even if envy is a "vice," as Rawls calls it, the irony is that if we lived in a world without it, there would be absolutely no reason to care about fairness and justice. There would never be any meaningful reaction to their absence, so why worry? Rawls's principles of justice sound eminently reasonable and might well help reduce envy, but isn't this precisely the point? In 1987 the German sociologist Helmut Schoeck wrote a whole book about envy, calling our species "man the envier." Without envy and attempts to forestall it, he claims, we couldn't possibly have built the societies we live in. Instead of denying this emotion, or regarding it as a threat to a well-ordered society, we should embrace and channel it. Schoeck urged us to "unmask" the role of envy in our lives, the way psychoanalysis unmasked the role of sex.[18]

Rational arguments are woefully insufficient to arrive at moral principles, which get their force from the emotions. The enormous investment we make in rectifying unfairness and injustice—the screaming protests, the marches, the violence, the endurance of police beatings and water cannons, the trolling and bullying on Facebook—remind us that we aren't dealing with some bloodless mental construct. Absence of fairness and justice shakes us to the core, something that no amount of elegant abstract reasoning will ever accomplish.

If a chimpanzee isn't treated the way he has come to expect, he will throw an ear-piercing tantrum, rolling on the ground in despair. It is overly dramatic but a great way to remind others to heed his wishes. As a result, in Taï Forest, in Ivory Coast, meat-sharing chimpanzees recognize one another's contributions to the hunt. Even the highest-ranking male is forced to beg and patiently wait for handouts if he comes late to the party. Clustered around the meat possessor, the hunters enjoy priority.[19] This is logical, because why would anyone help out if there were no link between effort and reward? It would be patently unfair not to share with hunters who have helped catch the prey. The emotional intensity of reactions to injustice is also well known in our own species, and explains why close-knit human societies disapprove of a winner-take-all mentality. Hunter-gatherers seem well aware of this mentality and actively discourage it, but modern society offers too many opportunities for individuals to flaunt it. The tendency to claim a disproportionate piece of the pie is so detrimental, however, that it even affects physical health.

Epidemiological data show that the more unequal a human society is, the shorter-lived its citizens are. Large income disparities tear the social fabric apart by reducing mutual trust, stirring up social tensions, and creating anxieties that compromise the immune system of both the rich and the poor.[20] The rich may retreat into gated communities, but that doesn't make them immune to the tensions. If inequality reaches extreme levels, a society may even face the explosive situation for which the French Revolution, for once, does carry an important lesson. Humans seek to level the playing field, and if efforts to achieve this are blocked for too long, we may bring in the guillotine.

I am still astonished that a simple experiment on capuchin monkeys led me down this path of speculation about fairness, one of humanity's most lionized moral principles. This was never my intention, but goes to show that one should always keep one's eyes peeled for unexpected behavior. The one-minute video of the cucumber-and-grape experiment that Sarah and I conducted went viral because people recognized themselves in the cage-rattling protest by the cucumber monkey. Some wrote to tell me they had forwarded the video to their boss to let him know how they felt about their salary. Others wrote of recognizing the same reaction in cable TV customers who heard that a neighbor with a new contract got a better deal. The monkeys themselves were unfazed by their fame.

When I closed my lab a few years ago, I was sad to see my monkey friends go but extremely happy that we found good homes for all of them. Half the colony went to the San Diego Zoo, where they became a popular attraction in a fantastic aviary-like area with tall climbing trees.

On a recent visit, my heart was warmed by how healthy and relaxed the monkeys looked, spoiled as they are by loving caretakers who know each one by name and keep them busy with spread-out foods and tool tasks. They told me that Lance, the cucumber thrower in the video, is still as quick-tempered as ever.

The other half of my colony stayed in science, moved to a forested facility in the Atlanta metro area where Sarah, now a professor at Georgia State University, continues exploring the limitations of the *Homo economicus* model, not just for our species but for primates in general. On my last visit, as I walked up to their outdoor enclosure, all the monkeys stayed on the ground. This was remarkable, because capuchins feel safer high above us. "They recognize you!" Sarah exclaimed. And that was before Bias, my favorite female, began flirting with me with friendly eyebrow flashes and head tilts, pointing in a direction behind her that suggested she knew a quiet spot.

Free Will and B.S.

In *Paradise Lost,* the seventeenth-century English poet John Milton felt that the fallen angels had too much time on their hands, so he sought to occupy them with a discussion topic. He chose free will. We all have the impression that we possess free will, although it lacks a clear definition and may be a total illusion. As Isaac Bashevis Singer once put it, "We must believe in free will, we have no choice." It is the perfect topic for eternal debate.

This debate relates to the emotions, because free will is often conceived as their opposite. Making a free rational choice requires us to deny or suppress our first impulses. In fact, the whole idea goes back to the debate over how much our mind is shaped by our body. Those who believe in free will argue that we can simply set aside the body and its nonvolitional desires and emotions and rise above them; humans— and humans alone—can fully control their choices and their destiny.

The opposite is a person without self-control, which philosophers have dubbed a *wanton*. A wanton follows whichever impulse hits first, whichever urge is most pressing and satisfying, and never looks back. Regret isn't something you'll find in a wanton. Young children and all animals are said to fall into this category.

We may capitalize Free Will to convey our reverence for a concept so central to human responsibility, morality, and the law, but if we can't measure it, how will we ever agree on it? Some say that free will boils down to making choices, but even bacteria make choices, and certainly all animals with brains have to decide between approach and avoidance, which prey to single out from the flock, or whether to travel north or south, and so on. The squirrels in my neighborhood decide whether to cross the road. Sometimes they do so right in front of my car, making me nervous. They run halfway, then quickly return, unable to make up their minds. Pairs of bluebirds in my backyard, getting ready to build their nest, visit every empty nestbox, hopping in and out multiple times, the male alternating with the female. They'd make excellent subjects for a *House Hunters* episode. After weeks of scouting, the male puts a few branches or grass stems into one of the boxes, then lets the female build the actual nest while he guards the site. The drawn-out decision process has reached its conclusion. Do bluebirds have free will?

Francis Crick, the British co-discoverer of DNA, proposed in his 1994 book *The Astonishing Hypothesis* that human free will resides in a very specific brain area: the anterior cingulate cortex. But humans are not the only species with this area, and we have good evidence that it also helps rats make decisions. Yet despite the signs that animals make choices every day, we refuse to grant that they have free will. Their choices are constrained by past experiences and inborn preferences, we argue, and animals lack the ability to fully review all the options in front of them.

Never mind that the same argument has been applied to great effect against free will in our own species, which is why some of history's greatest minds—Plato, Spinoza, Darwin—doubted its existence. Free will just doesn't fit the prevailing materialist worldview, as noted in 1884 by the prominent German evolutionist Ernst Haeckel:

> *The will of the animal, as well as that of man, is never free. The widely spread dogma of the freedom of the will is, from a scientific point of view, altogether untenable. Every physiologist who scientifically investigates the activity of the will in man and animals, must of necessity arrive at the conviction that in reality the will is never free, but is always determined by external or internal influences.*[21]

Of the plethora of definitions of free will, however, one strikes me as open for further investigation. The American philosopher Harry Frankfurt defined a "person" as someone who doesn't just follow his desires but is fully aware of them and capable of wishing them to be different. As soon as an individual considers the "desirability of his desires," he or she may be said to possess free will, Frankfurt asserted.[22] This is great, because it means that to test it, all we need to do is subject animals to a situation in which they would like to satisfy one desire but are also given a chance to refrain from action so as to satisfy another. Do they ever abandon their first desire?

They must be capable of doing so, because an animal that would give in to every impulse would constantly run into trouble. Being a wanton has no survival value. Migrating wildebeest in the Maasai Mara hesitate for a long time before jumping into the river they seek to cross. Juvenile monkeys wait until their playmate's mother has moved out of view before starting a fight. Your cat snatches meat from your kitchen counter only after you turn your back. Animals are keenly aware of the consequences of their behavior, which is why they often hesitate, like the squirrels in front of my car.

Sometimes they abandon a goal altogether, most noticeably in hierarchical systems. A young male chimpanzee who'd love to mate with a female approaches her and hangs around, hoping for an opportunity.

But when the alpha male looks in his direction, he will sneak off, knowing it's not going to work. Even more striking are the occasions when a high-ranking male comes around the corner and catches the young male spreading his legs to present his erection to the female—his not-so-subtle courtship signal. Upon seeing the high-ranking male, the young male quickly drops his hands over his penis, concealing it from view, well aware that he'd be in trouble if the other male so much as suspected what was going on. All this requires insight in what others know as well as the capacity to override one's urges. Aren't we getting close here to Frankfurt's definition of free will?

Frankfurt himself, however, leaves no room for free will in any organism other than the human adult, literally saying, "My theory concerning the freedom of the will accounts easily for our disinclination to allow that this freedom is enjoyed by the members of any species inferior to our own."[23]

This is B.S.!

Now, don't get me wrong, I don't like this kind of language any better than you, but since Frankfurt is the celebrated author of a 2005 book titled *On Bullshit,* I feel entitled to throw this word around. His book is a thoughtful, erudite treatment of the topic—with references to Wittgenstein and Saint Augustine—in which he explains in detail how bullshit compares to humbug, misrepresentation, and bluff. Bullshit is a creative overstatement that comes close to lying, which according to Frankfurt is "unavoidable whenever circumstances require someone to talk without knowing what he is talking about."[24] Frankfurt made his claim that species "inferior to our own" don't monitor their own desires without any indication that he knew what he was talking about, so it probably falls under the category of bullshit. It could also be bunk. Perhaps his claim about free will would have been reasonable when he first made it, fifty years ago, but new research contradicts it. We know a lot more about future orientation and emotional control in animals and young children than we did back then, and the situation is not nearly as simple as we once thought.

First of all, the popular notion that animals are captives of the present, that they live entirely in the here and now, has been blown up by recent work on "time travel." Apes, large-brained birds, and probably other animals as well think back to specific events in their lives and make plans for the future. Their minds travel in time. Chimpanzees sometimes collect handfuls of long grass reeds at one location in the forest, only to travel for miles to another location, carrying the reeds in their mouths, where they use them to fish for insects in termite hills. Most likely they had this goal in mind all along. Orangutan males give loud whooping calls, audible across wide stretches of the Sumatran rain forest, often from high up in a tree. I have stood underneath a calling male and can assure you that the calls leave you shaking. All the other orangs around listen carefully, because the dominant male (the only fully grown male with well-developed cheek pads) is a figure to be reckoned with. While building his nest for the night, he always calls in a specific direction, but a different one each night, corresponding to the direction he will set off in the next morning. This means he knows about twelve hours in advance where he'll be going, and he makes sure that everyone else knows as well.[25]

Other evidence for future orientation in animals comes from a string of controlled experiments in which primates or birds were presented with a tool or food that they could use or consume only the next day. Based on these studies, it is now widely believed that some animals possess a forward-looking cognition. Fairness studies are indicative, too. If chimps deliberately choose equal outcomes in the Ultimatum Game despite being aware that a different choice would yield more food for themselves, we need an explanation. The one I favor is that they sacrifice immediate benefits to preserve good relationships. If true, they are not only forward-looking but also endowed with excellent restraint.

Restraint is tested more directly with the marshmallow test. Most of us have seen the hilarious videos of children sitting alone behind a table desperately trying *not* to eat a marshmallow—secretly licking it, taking tiny bites from it, or looking the other way so as to avoid temptation.

Some animals control their emotions as well as humans do. In a classic test, children are promised that they will receive a second marshmallow if they refrain from eating the first. They struggle with the temptation, hide from the sugary treat, or distract themselves. Tested in a similar way, apes and one parrot, Griffin, held out for as long as children did. A bowl of food was placed in front of Griffin, who could get something better by waiting. The bird, too, often closed his eyes and invented distractions.

The children have been promised a second marshmallow if they leave the first one alone while the experimenter is away. The marshmallow test measures how much weight children assign to the future relative to immediate gratification. What do apes do when subjected to similar circumstances? In one study, a chimpanzee patiently stares at a container into which a candy falls every thirty seconds. He knows he can disconnect the container at any moment to swallow its contents, but he also knows this will halt the candy flow. The longer he waits, the more candies will collect in his bowl. Apes do about as well as children in this regard, delaying gratification for up to eighteen minutes.[26]

But what about, say, birds? Surely they don't need any self-restraint. But many birds pick up food that they could swallow themselves, yet carry to their hungry young instead. In some species, males feed their mates during courtship while going hungry themselves. Again, self-control is key. When Irene Pepperberg's African grey parrot, Griffin, was tested on a delayed gratification task, he managed long waiting times. As he sat on his little perch, a cup with a less preferred food, such as cereal, was placed in front of him, while he was asked to wait. Griffin knew that if he waited long enough, he might get cashew nuts or even candies. He was successful 90 percent of the time, enduring delays of up to fifteen minutes.[27]

The critical question in relation to Frankfurt's definition of free will is whether animals understand that they are fighting temptation. Are they aware of their own desire? When children avoid looking at the marshmallow or cover their eyes with their hands, we assume they feel the temptation. They talk to themselves, sing, invent games using their hands and feet, and even fall asleep so as not to have to endure the terribly long wait. The father of American psychology, William James, long ago proposed "will" and "ego strength" as the basis of self-control. This is how the behavior of children is typically interpreted. They are said to use conscious strategies to distract themselves.

The same may apply to apes. In the test with the falling candies, for example, apes hold out significantly longer if they have toys to play with. Focusing on the toys helps them take their mind off the candy machine. That they do so intentionally is indicated by the fact that they manipulate the toys a lot more during candy tests than otherwise.[28] Griffin the parrot, too, actively tried to block out the inferior food in front of him. About one-third of the way through one of his longest waits, he simply threw the cup with cereal across the room. On other occasions, he moved the cup just out of reach, talked to himself, preened himself, shook his feathers, yawned extensively, or fell asleep. He also sometimes licked the treat without consuming it, shouting "Wanna nut!"

Given the striking similarity in behavior between children, apes, and Griffin, it is best to assume shared mental processes, including awareness of one's own desires and deliberate attempts to squelch them. And so the answer to the perennial question of free will is that if we assume it for ourselves, we probably also need to grant it to other species. Otherwise, it's unclear what we should make of all the inhibitions animals exhibit both in experimental settings and in the wild.

Take the case of a mother chimp whose infant is picked up by a well-meaning female adolescent. This is a daily scenario, as young females are irresistibly drawn to babies and always want to hold and cuddle them. Unfortunately, they are also clumsy. The mother knows this and will follow the adolescent, whimpering and begging, trying to get her offspring back. The adolescent keeps evading her, however. The mother suppresses an all-out pursuit for fear that the kidnapper will escape into a tree and endanger her precious baby. For the same reason, she cannot simply grab her baby. Imagine two females, each pulling at a limb, stretching a screaming infant between them! I've seen it happen, and it's a most disturbing sight. So Mom needs to stay calm and collected. She may even act as if she's hardly interested, sitting nearby with a casual air, munching on grass or leaves, just to show that she poses no threat. Once the infant is safely back clinging to her belly, however, everything changes. I have seen such a mother turn on the adolescent, chasing her over long distances with furious barks and screams, releasing all her pent-up fury. The whole sequence gives the impression that the mother held her extreme worry and irritation in check for the sake of a safe outcome.

As I've already mentioned, subordinate primates have to suppress or hide their desires in the presence of dominants, but the opposite is also true. A field experiment in South Africa put this to the test by training a low-ranking female in a group of wild vervet monkeys to become an expert. Called the provider, she was the only one who knew how to open a food container, but she was smart enough to do so only if there were no dominant individuals nearby to steal her food. So she'd wait to perform her trick until all higher-ups were at a safe distance. The dominants, in turn, learned what distance they should keep from the container for there to be any chance that the provider would open it.

After many repeated trials in three different monkey groups, the investigators reported that the dominants showed incredible patience and discretion. They remained outside an imaginary "forbidden circle" of about ten meters around the container, often waiting in a distant tree from which they could keep an eye on it. Once the provider opened the container, she would grab food with both hands and quickly stuff it into her cheek pouches, a handy attribute of ground-dwelling vervets. Once her cheeks were bulging with peaches, apricots, and dried figs, she didn't mind being displaced when others hurried over to follow her example. She simply moved to a quiet spot to consume at leisure everything she'd collected. Without the higher-ups' self-imposed restraint and queuing, this whole operation would never have worked to everyone's advantage.[29]

There are lots more examples of self-restraint. Anyone who has a large and a small dog at home can see it in action whenever they play together. One of the most remarkable expressions of self-control is toilet training. Canines have a natural basis to relieve themselves outside the den, while felines deposit waste in dirt where it can be covered. Our domesticated pets still have to be trained, but their natural tendencies are a tremendous help. For children, toilet training is their first step toward control over bodily functions and self-control in general. Freud made a big deal about it, depicting it as a fierce battle between the id, which seeks the pleasure of relief, and the superego, which absorbs society's rules and wishes.

For apes, on the other hand, you'd think that toilet training would never work even though they are so similar to us. Wild apes travel through trees and build a different nest every night, showing little concern about when and where to pee or poop—it just drops to the ground far below them. Nevertheless, toilet training has been tried on home-reared apes.

In the 1930s, Winthrop and Luella Kellogg raised the young chimpanzee Gua at home, and took a total of nearly six thousand notes about her potty training. The Kelloggs kept the same notes for their son, Donald, so that they could compare both "organisms," as they called them, in both urination and defecation. At the beginning, the ape learned more slowly, but remarkably, after about one hundred days, the two were equivalent in the number of evacuation errors they made, which kept dropping. Gua and Donald reached the final stage when they were both able to timely announce their impending discharge, which they both started doing at around one year. Their typical signal was to press their hands firmly against their genitals. The only difference was that Gua also did so with her foot, hobbling grotesquely around on two hands and the remaining foot. Approaching her adoptive parents this way, she'd vocalize, then later announce her needs by just calling out to them. I find this a most impressive act of willpower in a species that normally wouldn't need to exercise it in relation to bodily functions.[30]

Animals just can't afford to blindly run after their impulses. Their emotional reactions always go through an appraisal of the situation and judgment of the available options. This is why they all have self-control. Furthermore, in order to avoid punishment and conflict, the members of a group need to adjust their desires, or at least their behavior, to the will of those around them. Compromise is the name of the game. Given the long history of social life on earth, these adjustments are deeply ingrained and apply equally to humans and other social animals. So even though personally I am not a big believer in free will, we do need to pay close attention to the way cognition may override inner urges. Fighting the impulse to take one course of action and replacing it with another that promises a better outcome is a sign of rational agency. It is also essential to any well-ordered society, which is why the American psychologist Roy Baumeister remarked, "Perhaps ironically, free will is necessary to enable people to follow rules."[31]

I suggest, therefore, that we expand the perennial debate about this issue by asking why it is customarily assumed that free will makes us human. What exactly is it about us that makes us so sure that we have it and other species don't? Why do we think we're the only ones with the freedom to determine our future? Given the above evidence, the reason for the presumed difference can't be control over our emotions and impulses, or even awareness of our own desires. I'd love an answer that we can put to the test, because we'll never get there with the prejudices that have informed the debate thus far. Until then, my tentative conclusion is that if we humans did evolve free will, it is unlikely that we were the first ones.

Stand by Me

Now that we are finally allowed to talk about animal emotions, in our relief we tend to forget how little we know. Compared with psychologists working on humans, we are light-years behind. We name a couple of emotions, describe their expression, and document the circumstances under which they arise, but we lack a framework to define them and explore what good they do. Or perhaps we are not so far behind, because human emotions lack such a framework, too. Biologists always think in terms of survival and evolution, so it is logical for us to ask how emotions affect behavior. We care more about the action side than the feeling side because the value of emotions is in the behavior that they generate, from a baby's hungry cry to an elephant's annoyed charge. Emotions evolved for a reason, and whereas natural selection cannot "see" feelings, it does pay attention to actions with consequences. Yet, how the emotions evolved remains a mystery.

An even greater mystery is how emotions are regulated to guarantee optimal outcomes. Emotions don't always know what's best for the organism. Most of the time they do, but sometimes it's better to ignore them or steer our behavior in a different direction. We humans employ fancy terminology to describe how we handle this problem, such as *executive function, effortful control,* and *emotion regulation.* These capacities are crucial to how we plan and organize our lives. But we almost never apply this terminology to animals, based on the prejudice that animals have few emotions and are unable to disobey them. But not only do animals show self-control in tests of delayed gratification, they also often face conflicting emotions that pull them into opposite directions. They hesitate between fight and flight, between weaning their offspring and giving in to their tantrums, between avoiding an attacker and reconciling, or between copulating and repelling a rival.

One of my male students had the misfortune of being regarded as competition by an adolescent male chimpanzee named Klaus. Every time this student walked by, Klaus would display and throw mud or body waste at him, expressing profound dislike. Klaus never did so with me or any other people. In fact, we considered him sweet and playful. One day Klaus was courting a female in the outdoor area, and precisely at the moment that he got lucky and the female answered his call, his human enemy showed his face. Klaus abandoned the female and went straight into an angry display. The lure of sex was no match for his desire to show off, with all his hair on end. Perhaps he had reached the stage in life at which he absolutely needed to prove his place in the hierarchy, and who better to pick on than someone of similar age and sex albeit of a different species? Klaus may have calculated that his tryst with the female could wait.

We need to start paying attention to such calculations, which humans make all the time. We deftly navigate our emotions and desires, following some, resisting others. We set priorities in order to arrive at the best decision, a remarkable capacity often attributed to our cerebral cortex. We're told that humans owe their high foreheads to the exceptional size of this part of the brain, which is the seat of higher cognition and impulse control. We consider our foreheads "noble" and even have a long sordid history of comparing them between races (as in, for example, the "Aryan forehead"). Yet foreheads say remarkably little about skull content, and the human brain doesn't structurally differ much from those of monkeys and apes. Relative to the rest of the brain, our cerebral cortex is unexceptional. The latest neuron-counting techniques back this up. In humans, the cerebral cortex holds 19 percent of all neurons in the brain, the same as in other primates. Brains of humans and apes start out at similar size in the fetus, but then the human brain keeps expanding throughout gestation, whereas the ape brain's growth slows down about halfway.[32] The result is an adult human brain that is three times larger and counts more neurons (a total of 86 billion) than any ape brain. We may not have a different computer, but we definitely have a more powerful one. No one is saying that human cognition isn't special, but it's time to recognize that the interplay between intelligence and emotions, as reflected in frontal lobe dimensions, is likely to be the same across all the primates.[33]

A great deal of emotion regulation takes place unconsciously and is part of social relations. This is why I have a problem with the way psychologists generally test human emotions: they put subjects alone in a chair behind a computer or alone in a brain scanner, despite the fact that most of our emotions evolved in social settings. Emotions are not individual but interindividual. The American neuroscientist Jim Coan took a different approach by testing human subjects in a scanner to measure neural responses to a signal that announced the arrival of a mild electrical shock. The brain images showed, as one would expect, that people worried about the upcoming pain. But then Coan added a second person—female participants were allowed to hold the hand of their husband—and found that the fear dissipated, and the upcoming shock seemed only a minor irritant. Moreover, the better the woman's relation with her husband, the more effective was the buffering. The reverse wasn't tested, but it probably would have produced the same outcome.

Another study found that handholding synchronized the brain waves between both partners. This is a powerful demonstration of how attachment and body contact modify emotional reactions.[34]

After attending a lecture by Coan, I complimented him on his experimental design. He told me that most psychologists believe that our species's typical responses occur while we are alone. They regard the solitary human as the default condition. Coan, however, believes the exact opposite: how we feel while we are embedded with others is the actual norm. Few of us deal with life's stresses on our own—we always rely on others. In experiments, women are less easily stressed if they can smell a T-shirt worn by their husband or romantic partner. The reassuring effect of this familiar scent might explain why people, when home alone, often wear their partner's shirt or sleep on their side of the bed.[35] Western culture has a love affair with autonomy, but in our hearts and minds we are never truly alone. As biologists know, humans are obligatorily social (we can't survive outside a group and we suffer mentally if held in isolation); hence our norm is the way we function in a social milieu, with all the emotional cushioning that it entails. It's not so different from the way my capuchin monkeys contact-call nonstop. Even when apart, these monkeys consider themselves part of a group and constantly seek reassurance that everyone is still there for them. They are vocal handholders.

The deepest disruption of emotional life occurs when individuals are deprived of an affectionate environment while growing up. We are not made to fend for ourselves, and neither are any other primates. The first time I examined how upbringing affects the emotions was during a study of bonobos at the Lola ya Bonobo Sanctuary, near Kinshasa. Sadly, all these bonobos are traumatized orphans. Poachers and hunters routinely kill wild bonobos (along with many other animals) for bushmeat, and any bonobo youngster found clinging to a dead victim is "rescued" and sold alive. Because this is against the law, these baby apes are confiscated from the market and brought to the sanctuary, where they are raised by *mamans*—local women who watch over them, carry them around, and give them the bottle. After a couple of years, they join the colony in a fenced-in forest where they await the moment, years later, when they may be released back into the wild.

My co-worker Zanna Clay set out to study levels of empathy among the orphaned bonobos. One index of empathy is the way bystanders react to distress caused by a fight: they may wrap both arms around the screaming loser of a confrontation and console them by gently holding and stroking them; they may even walk away with them with an arm around their shoulders. These acts calm down losers, who may stop screaming with amazing abruptness.

Whenever a spontaneous fight broke out in the large bonobo colony, Zanna recorded it on video so we could analyze it in detail. We observed moderate levels of empathy in the orphans. But to our astonishment, the real compassion champions were the half-dozen bonobos who had been born in the colony and been raised by their own mothers. Bonobos with this background proved far more inclined to comfort others who were in pain or distress. Taking their behavior as the norm, we concluded that being an orphan seriously hurts an individual's ability to empathize.[36]

We know how important emotion regulation is for human children. In order to show empathy, they need to control their own distress. A young child who sees and hears another child cry may become distressed herself, resulting in two crying children. The second child is not as deeply distressed as the first, however, and often snaps out of it. This allows her to pay attention to the first and provide comfort. Children who are unable to downregulate their own emotions, however, are overwhelmed and not good at showing concern for others.[37]

Empathy may work the same way in bonobos. Orphans have trouble self-regulating, whereas mother-reared bonobos do not, having learned how to modulate emotional upheavals. Zanna tested this idea by observing how individuals handle their own distress. Orphans, she found, are slower to shift from one emotional state to another, and remain upset for longer periods than do mother-reared bonobos. The ones who keep screaming and screaming after they have been rebuffed or bitten by another are the same ones who rarely comfort others. It's almost as if an individual first needs to have his own emotional house in order before he is ready to visit someone else's emotional house. The orphans' deficit is fully understandable as they have suffered unimaginable abuse at the hands of humans, losing their mothers to poacher snares or bullets at a tender age. Poachers also may have kept them chained to a tree for months. It is remarkable that they show any empathy for their fellow bonobos at all.

This work has taught me that in addition to studying the emotions of animals, we should also explore how animals manage them. That may reveal key differences between species as well as between individuals, defining their personalities. Self-regulation is a rich topic that has also been applied to human orphans, such as those discovered in Romania following the overthrow of Nicolae Ceauşescu in 1989. The world was shocked by the conditions in which these children lived. One British journalist, Tessa Dunlop, reported: "When I first walked into the large grey building at the heart of Siret, my immediate instinct was to walk straight back out again. Half-naked children leapt from every direction, clawing at my clothes, and there was an overpowering smell of urine and sweat that made me want to retch."[38] These orphans grew up without love or affection, under supervisors who abused them and incited violence, such as by asking older children to beat up younger ones. We know from brain research that institutionally reared children have enlarged, overly active amygdalas—an area of the brain involved in emotional processing—and pay excessive attention to negative information. They are easily frightened. Their emotion regulation and mental health are permanently damaged, which is why the Romanian orphanages were known as the "slaughterhouses of souls."

There are many parallels with animals reared in isolation, such as the dairy industry's horrific practice of separating calves from their mothers right after birth. It inflicts deep emotional disturbance upon both the cows and their offspring. These calves are less socially active and adept, and more easily stressed, than those that have been allowed to stay with their mothers: their emotional appraisals are messed up, and they are quickly thrown off balance.[39] We know far too little about these processes, partly because of the long-standing taboo on animal emotions, but also due to the popular notion that animals are mere wantons without emotional control. For cows, bonobos, and many other species, however, emotional intelligence is absolutely crucial. Their boats don't just drift down a river of feelings—they are equipped with rudders and oars to help them navigate. Being raised without love and attachment removes these tools, which is why orphans have such trouble achieving emotional equilibrium.

7 | # SENTIENCE

What Animals Feel

> *"Of course, what I felt then as an ape I can represent now only in human terms, and therefore I misrepresent it."*
>
> Franz Kafka (1917)[1]

When people ask me if I think an elephant is a conscious being, I sometimes retort, "You tell me what consciousness is, and I'll tell you if elephants have it." This usually shuts them up. No one knows exactly what we're talking about.

My answer is unfair, though, and a bit mean, both to the questioner and to elephants, because I do in fact grant these lumbering giants consciousness. When my team worked with Asian elephants, we were the first to show that they recognize themselves in a mirror, which is often viewed as a sign of self-awareness.[2] We tested their cooperative skills, such as how well they understand that they need a helping trunk for a joint task. Elephants did as well as apes and better than most animals. Everything about them strikes me as deliberate and smart. For example, when young elephants in Thai or Indian villages are outfitted with bells around their necks to announce their whereabouts (and keep them from surprising people in their gardens or kitchens), the animals sometimes stuff grass inside the bells to muffle the sound. That way they can walk around undetected. This solution suggests imagination, because certainly no one showed them how to do this, and grass doesn't accidentally get inside bells for them to discover its effect. To come up with clever solutions, we humans consciously put cause and effect together in our heads. If this is how we do it, why would elephants have a shortcut to problem solving *sans* consciousness?

Once at a symposium, I heard a prominent philosopher say he was going to explain human consciousness, claiming it was the logical outcome of our huge number of neurons. The more neurons interconnect, he said, the more conscious we become. He even played a video of the growth of a dendrite, which was amazing to watch but did nothing to help me grasp how consciousness arises. His most surprising conclusion was that human consciousness is off the scale compared to that of any other species. We are by far the most conscious of all creatures, he said, as if this were natural. But I didn't see how this conclusion followed from his theory about neurons and synapses, given that we are not the only ones with lots of them. What about animals that have brains larger than our 1.4-kilogram ones, such as the sperm whale's 8-kilogram brain?

But all right, I thought, humans possess more neurons, so perhaps his theory holds up. It had always been taken for granted that our brain has the most neurons—until we started counting them. It has now come to light that the elephant's 4-kilogram brain actually has three times as many neurons as ours does.[3] This discovery has caused a lot of head-scratching. Do we need to rewrite the story of human consciousness? What proof is there actually that we are more conscious than elephants? Is it only because they don't talk? Or because most of their neurons are in a part of the brain that isn't associated with higher functions? This last one seems a good argument, except that we don't know exactly how each part of the brain contributes to consciousness. This animal has a three-ton body and forty thousand muscles in its trunk alone (not to mention its prehensile penis); it must carefully orchestrate every step (think of the tiny calves walking between the legs of their mothers and aunties), and it has more genes dedicated to smell than any other species on earth. Are we certain that it is less aware of its own physique and its surroundings than we are? The complexity of a body, its moving parts, and its sensory inputs are no doubt where consciousness began. In this regard, the elephant is second to none.

Not all philosophers agree with the one who postulated that consciousness requires a huge brain. With the rise of animal studies and anthrozoology (the study of human-animal interactions), many open-minded philosophers have begun to think about animal sentience in ways that invite further research. They recognize that even if we may never know *what* elephants feel, we may still be able to establish *that* they feel.[4] Without a clear understanding of what consciousness is, how can

One elephant consoles a distressed companion by wrapping her trunk around her while rumbling. Elephants are highly empathic, emotional beings, but what they feel remains unknown to science. Since it has been argued that feelings and consciousness depend on the number of neurons in one's brain, the recent discovery that elephants have three times as many neurons as humans has shaken preconceptions.

we exclude this possibility? Anyone who tries to resolve this issue by telling you that there are many different kinds of consciousness (self-awareness, existential consciousness, body awareness, reflective consciousness, and so on) is compounding the problem by adding foggy distinctions to a concept that is already foggy enough.

It is with some trepidation, therefore, that I wade into the morass of animal sentience and consciousness.

Meat and Sentience

Behind the debate about animal consciousness lurks an issue that many scientists would much rather avoid: what humanity does to animals. Clearly, we don't treat them well, at least not most of them. It's easier for us to live with this fact by simply assuming that animals are dumb automatons devoid of feelings and awareness, as science has done for a long time. If animals are like rocks, we can throw them onto a heap and stomp on them. If they are not, however, we have a serious moral dilemma on our hands. In this era of factory farming, animal sentience is the elephant in the room. There are thousands of animals in zoos, millions in labs, and millions more in human homes, but literally billions and billions of animals in the farm industry. Of the entire terrestrial vertebrate biomass on earth, wild animals constitute only about three percent, humans one-quarter, and livestock almost three-quarters!

On old-fashioned farms, animals had names, pastures to graze in, mud to wallow in, or sand to dust-bathe in. Life was far from idyllic, but it was appreciably better than it is nowadays when we lock up calves and pigs in narrow crates of stainless steel, cram chickens by the thousands into sunless sheds, and don't even let cows graze outside anymore. Instead, we keep them standing in their own waste. Since these animals are mostly kept out of sight, people rarely get to see their miserable conditions. All we see is cuts of meat without feet, heads, or tails attached.

This way we don't need to ponder the meat's existence prior to packaging. And here I am not even talking about the fact that we eat animals, only about how we treat them, which is my main concern.

I am too much of a biologist to question the natural circle of life. Every animal plays its role by eating or being eaten, and we are involved at both ends of the equation. Our ancestors were part of a vast ecosystem of carnivores, herbivores, and omnivores, ingesting other organisms and also serving as meals for predators. Even if nowadays we rarely fall prey anymore, we still let hordes of critters devour our rotting corpses. It's all dust to dust.

Our closest primate relatives go out of their way to catch monkeys and duikers in the forest, which they do with daredevil skills while employing high-level collaboration. They feast on their prey with great relish and happy grunts. They also spend hours fishing with twigs for ants and termites. Some chimp populations consume large quantities of animal protein (in one forest, they nearly wiped out the red colobus monkeys), whereas other populations survive on less.[5] Male chimps can double their mating success by using meat to buy sex.

Humans, too, prize meat immensely and eat it whenever they get a chance. We may not have fangs and claws like specialized carnivores, but we do have a long evolutionary history of supplementing a diet of fruits, vegetables, and nuts with proteins from vertebrates, insects, mollusks, eggs, and so on. And not just supplementing: according to the latest anthropological research, 73 percent of hunter-gatherer cultures worldwide derive more than half of their subsistence from animal foods.[6] This omnivore background is reflected in our multifunctional dentition, our relatively short intestines, and our massive brains.

Attraction to meat has shaped our social evolution. The gathering of fruits, which are small and dispersed, is mostly an individual job, but the hunting of large game demands teamwork. One man alone doesn't bring home a giraffe or mammoth. Our ancestors deviated from the apes by hunting animals larger than themselves, which required the sort of camaraderie and mutual dependence that is at the root of complex societies. We owe our cooperative nature, our food-sharing tendencies, our sense of fairness, and even our morality to the subsistence hunting of our ancestors. Furthermore, since carnivores are on average larger-brained than herbivores and since brains require a great deal of energy to grow and operate, the consumption of animal protein along with effective food processing (such as fermentation and cooking) are seen as driving forces behind our ancestors' neural expansion.[7] Animal protein provided them with the optimal mix of calories, lipids, proteins, and essential B_{12} vitamins to grow large brains. Without meat, we might not have become the intellectual powerhouses we are today.

None of this is to say, however, that we have to keep eating the way we do, or even eat meat at all. Animal protein may be overrated. We live in different times with different possibilities, and we have promising alternatives in the works, such as *in vitro* and plant-based meats that can be stuffed with all the vitamins we need.

Even if I have no problem with meat-eating per se, there is a lot wrong with how we treat animals and how we raise, transport, and slaughter them. The conditions are often degrading and sometimes plainly cruel. In reaction, many young people in the industrialized world are experimenting with meatless diets, even though this regimen remains challenging. A 2014 study by the U.S. Humane Research Council found that only one out of every seven self-declared vegetarians maintains their diet for over a year.[8] I admire the effort, though, and have joined it in my own imperfect and undogmatic fashion by banishing practically all mammalian meat from my family's kitchen.

Trends such as flexitarianism (a semivegetarian diet with occasional meat) and reducetarianism (reduced meat consumption) enjoy great momentum. A plant-based food revolution is under way that hopefully will force meat producers to change their methods. It would be great if humanity could cut its meat consumption by half while drastically improving the lives of the animals that it does eat. Maybe we could even go further and phase out actual animals altogether by producing meat separate from a central nervous system in a petri dish. I see the pursuit of such goals as a moral imperative, but it will be best accomplished if we honestly face where we come from rather than spinning the fairy tale, often heard these days, that we are meant to be vegan. We are not.

As a result of these ongoing debates, *sentience* has become a term that is both popular and loaded. It is one of three reasons (apart from pressing ecological ones) that humans should respect all forms of life: the inherent *dignity* of all living things, the *interest* every form of life has in its own existence and survival, and *sentience* and the capacity for suffering. Let me go over these reasons one by one. They pertain to all organisms regardless of whether we classify them as animals, plants, or something else.

Nothing requires us humans to assign dignity to particular organisms, so it is up to us to do so. Perhaps it shouldn't be this way, but this is how we operate. I may not assign any dignity to a mosquito in my bedroom or to a weed in my garden, but I realize that these are self-serving choices. I have more respect for a stunning butterfly or a cultivated rose. Clearly, the way we assign dignity is subjective. The only objective criteria might be an organism's intelligence and its age. We generally hold animals with large brains in higher regard than those with small ones, even though this, too, may be a human prejudice given that we are large-brained ourselves. We are equally biased toward our fellow mammals. We rate a dolphin higher than a crocodile, and a monkey higher than a shark. I am always suspicious of these judgments, though, as they fit too nicely with the old *scala naturae,* which lacks scientific standing. As for age, we admire longevity. I have multiple white oaks around my home in Georgia, some of which may be over two hundred years old. I assign great dignity to these tall trees, as I do to all individual organisms of a certain age, such as an old elephant, tortoise, or lobster. Some towns in Europe have a thousand-year-old linden tree in a central spot on the market square, often appropriately named *Lindenplatz.* No one with any respect for nature would lightly consider removing such a beautiful tree. It would be like bulldozing a cathedral.

Interest in staying alive is the easiest of the three reasons for respecting life on this planet, because it marks each and every organism. All life-forms do their best not to be eaten by hungry enemies, and all seek to acquire enough energy to survive and reproduce. They may not do so consciously, but clinging to life is part of being alive—no exceptions. Even single-celled organisms rapidly swim away from a toxic substance. Plants release toxic chemicals to repel enemies and warn each other chemically via the air, or through their root systems, about external threats such as grazing bovines or munching insects. The survival interests of organisms often collide, so that one organism cannot survive without infringing upon another's interests. This is certainly true for all animals, which lack the capacity to convert sunlight into energy. As a result, animals must ingest organic matter to obtain the calories they need for survival. All animals maim or kill other organisms. Even the most organic vegetable farmer cannot help but violate the interests of other life-forms by stealing the habitat of wild animals, eradicating insects with natural pesticides, and sacrificing plants to human consumption. Being part of the fabric of nature, we constantly weigh our interests against those of other organisms, typically favoring our own.

The sentience question is the most complex of the three. Sentience is defined as the capacity to experience, feel, or perceive. In its broadest sense, sentience characterizes every organism, such as the eukaryotic cell, which strives for a steady chemical balance inside its walls. Seeking homeostasis requires the cell to sense its interior concentration of oxygen, carbon dioxide, pH level, and so on, and to "know" what actions, such as osmosis, to undertake to restore balance. The American microbiologist James Shapiro has gone so far as to claim that "living cells and organisms are cognitive (sentient) entities that act and interact purposefully to ensure survival, growth, and proliferation."[9] Similarly, the neuroscientist Antonio Damasio writes about the cell in *The Feeling of What Happens,* a 1999 book about inner experiences:

> *It requires something not unlike perception in order to sense imbalance; it requires something not unlike implicit memory, in the form of disposition for an action, in order to hold its technical know-how; it requires something not unlike a skill to perform a pre-emptive or corrective action. If all this sounds to you like the description of important functions of our brain, you are correct. The fact is, however, that I am not talking about a brain, because there is no nervous system inside the little cell.*[10]

This broad meaning of sentience applies equally to plants. Even though plants move extremely slowly, which makes their "behavior" hard to detect, they do sense changes in their environment (light, rain, noise) and take measures to fight threats to their own existence. For example, thale cress (a small flowering weed related to broccoli) defends itself against insects by producing toxic mustard oils in its leaves, doing so more when scientists play the vibrations of chewing caterpillars than those of birdsong.[11] Plant "behavior" can be quite complex, such as the heliotropism of sunflowers, which track the sun's movement across the sky yet reorient overnight to the east where the sun will come up. I put "behavior" between quotation marks, however, because it mostly boils down to the release of chemicals and directional growth, even if some plants respond faster, such as the carnivorous Venus flytrap, which closes its leaves around insects, or plants that respond to touch, such as mimosa. In a curious parallel to the loss of consciousness that mammals are capable of, these plants lose their touch-sensitivity and mobility in response to the same medical anesthetics applied to hospital patients.[12]

Science has only scratched the surface of the sophisticated defenses, alarm signals, and mutual support systems of plants, which surely suggest that they "don't like to be eaten," as it is sometimes put. It goes too far, however, to claim that by releasing gases upon attack, plants "cry out in pain." It is fine to speak of active resistance to threats and striving for survival, but in order to feel pain, plants would need to experience their own condition. While there are electrical pathways inside plants that resemble animal nervous systems, no one knows if stimulating these pathways induces subjective states, especially since there is no brain to register and ponder them. For most scientists, consciousness in the absence of a brain is a nonstarter. This is where we run into limitations of the sentience label for plants. They may very well react to their environment, and maintain an inner balance of fluids, nutrients, and chemicals, without feeling anything at all. Reacting to environmental changes is not the same as experiencing them.

Sentience in the narrow sense implies subjective feeling states, such as pain and pleasure. If we doubt that plants feel anything and deny them this kind of sentience, we should do the same for animals without a central nervous system. We don't know if oysters and mussels, for example, experience internal states given that they have only a few nerve cords and ganglia (clusters of nerves) and no brain. Like plants, they have no (oysters) or limited (mussels) ways of avoiding painful situations by moving out of their way. Apart from clamming up, they lack the behavioral apparatus for which pain sensations would make sense. I am reluctant to grant bivalves sentience in the narrow sense, therefore.

But whatever our opinion, we should be consistent and either consider both plants and bivalves sentient or else deny sentience to both of them as well as to all other organisms without brains, such as fungi (a very interesting group that is neither plant nor animal), microbes, sponges, jellyfish, and so on. That these organisms belong to different taxonomic kingdoms is immaterial given that all organic life is built on the same principles. At the same time, it will serve us well to recall the long history of science in underestimating animals. At this point, there is no guarantee that we aren't doing the same with plants.

When we get to species equipped with brains, sentience becomes far more likely. Everyone readily believes in it for elephants, apes, dogs, cats, and birds, but we should also consider species with smaller brains. Inside the lab of Barry Magee and Robert Elwood at Queen's University in Belfast, shore crabs were offered dark locations to hide from bright lights. As soon as they entered some of these hideouts, however, they received electric shocks. The crabs soon learned to avoid these particular spots. This went beyond a reflex-like aversion—similar to the way plants chemically deter predatory insects—because it required the crabs to recall the precise context in which they had been shocked. Why would they change their behavior if they had not had a memorable experience? They must have actually *felt* the pain. The issue is even more complex, because experiments on hermit crabs found that when they have a particularly nice shell to protect their abdomen, they need a higher shock level to evacuate the shell than if they own a poor-quality shell. Apparently hermit crabs weigh aversive experiences against the advantages of a suitable home.[13]

If arthropods feel pain, as these experiments imply, we should consider them sentient in the sense of having subjective feeling states. This includes the lobsters that we boil alive and the insects that we exterminate by the trillions. Whether these states resemble ours or those of mammals in general is not the issue. What matters is that these animals feel and remember. By extension, I'd suggest applying this rule to all animals with a central nervous system unless we find evidence to the contrary. Thus, I was baffled to learn that scientists at the Salk Institute in California, who are creating human-pig chimeras (a cellular mix of both species), are desperate to keep these man-made organisms from "becoming sentient." They want to block human cells from settling in the host brain to prevent the chimera from having a human mind.[14] Not only do these scientists overestimate what a few runaway human cells can accomplish, but they also overlook that pigs by themselves are already plenty sentient.

Chrysippus's Dog

Chrysippus, a Greek philosopher from the third century BC, is said to have recounted how a hunting dog arrived at a spot where three roads met. The dog smelled the two roads by which the quarry had not passed, then without hesitation or any further sniffing set off on the third. According to the philosopher, the dog had drawn a logical conclusion, reasoning that if the quarry had not taken two of the roads, it must have taken the third. Great thinkers, and even King James I of England, have used Chrysippus's dog to argue for the possibility of reasoning in the absence of language.

Facing a fork in a maze, mice often hesitate for a few seconds before continuing. Recent studies suggest that in order to decide which way to go, a mouse has to project itself into the future. We know that rodents replay previous action sequences in their hippocampus. So the wavering mouse in the maze probably compares the memory of old routes with imagined future ones. In order to do so, he will have to be able to tell the difference between experienced and projected actions, which requires a primal sense of self. At least, this is what the scientists doing these experiments assume. I find this fascinating, because in this thought experiment, we postulate that humans would need a sense of self to take the same decision, which we then take as evidence for a sense of self in another organism. This extrapolation is generally satisfying, but not risk free, because it hinges on the assumption that there is only one way to solve a problem.[15]

Chrysippus's dog is a great example of apparent *inferential reasoning*. For me, the main question is not the role of language, but whether this kind of reasoning implies consciousness. Fortunately, we now have tests of inferential reasoning. The American psychologists David and Ann Premack presented their chimpanzee, Sarah, with two boxes, putting an apple in one and a banana in the other. After a few minutes, Sarah would watch one of the experimenters munch on either an apple or a banana. This experimenter then left the room, and Sarah was given a chance to inspect the boxes. She faced an interesting dilemma, since she had not seen how the experimenter had obtained his fruit. She couldn't be sure if it had come from the boxes. Invariably, however, she would go to the box with the fruit that the experimenter had *not* eaten. She must have concluded that the experimenter had taken his fruit from the corresponding box and that the second box would still contain its original fruit. Most animals don't make any such assumptions, the Premacks note; they just see an experimenter consume fruit, that's all. Chimpanzees, by contrast, always try to figure out the order of events, looking for logic, filling in the blanks.[16]

In another test, apes were presented with two covered cups after they had learned that only one would be baited with grapes. Both cups were covered and shaken. As expected, the apes preferred the cup in which they could hear the grapes make noise. But then the experimenter shook just the empty cup, which obviously made no sound. The apes picked the other cup. Based on the absence of sound, they inferred where the grapes must be,[17] much like the dog that was said to take the third road based on lack of scent on the other two.

I once watched another such causal inference unfold at the Burgers Zoo when the chimpanzees in the indoor colony watched us carry a crate full of grapefruits, which they find delicious, through a door that went outside onto their island. They seemed interested enough. But when we returned to the building with an empty crate, pandemonium broke out. As soon as they saw that the fruit was gone, twenty-five apes burst out hooting and hollering in a most festive mood. Like children waiting for an egg hunt, they must have inferred that the grapefruits were on the island where they'd soon spend their day.

Animal consciousness is hard to investigate, but we are getting close by exploring examples of reasoning, such as those given above, that we humans cannot perform unconsciously. We cannot plan a party without consciously thinking about all the things we need; the same must apply when animals plan for the future. The latest neuroscience suggests that consciousness is an adaptive capacity that allows us both to imagine the future and to connect the dots between past events. We are said to have a "workspace" in the brain where we consciously store one event until another one comes along.[18] Take for example taste aversion in rats. It is well known that rats avoid certain toxic foods even if they don't become nauseous until hours later. Simple association fails to explain this.[19] Could it be that rats consciously go over the recent past in their minds, thinking back to every food encounter to determine which one was most likely to have made them sick? We certainly do so ourselves after food poisoning and gag at the mere thought of the particular food or restaurant that we believe caused a shock to our digestive system.

The possibility that rats have a mental workspace where they review their own memories is not so far-fetched given the growing evidence that they can "replay" memories of past events in their brain.[20] This kind of memory, known as *episodic memory,* is different from associative learning, as when a dog learns that by responding to the command "sit," he will be rewarded with a cookie. To create the association, the trainer has to give the dog the reward right away—an interval of even just a few minutes is not going to be helpful. In contrast to this kind of learning, episodic memory is the capacity to think back to a specific event, sometimes long ago, the way we do when we think of, say, our wedding day. We remember our clothes, the weather, the tears, who danced with whom, and which uncle ended up under the table. This kind of precise memory requires consciousness, the way Marcel Proust, in *Remembrance of Things Past,* relived his childhood after tasting a little tea-soaked madeleine. Colorful and alive, these memories are actively called up and dwelled upon.

Episodic memory must be at work when foraging wild chimpanzees visit about a dozen fruit-bearing trees each day. The forest has far too many trees for them to go about it randomly. Working in Taï National Park, in Ivory Coast, the Dutch primatologist Karline Janmaat found the apes to have an excellent recall of previous meals. They mostly checked trees from which they had eaten in previous years. If they ran into copious ripe fruit, they'd gorge on it while grunting contentedly, and made sure to return a few days later. Janmaat describes how the chimps would build their night nests en route to such trees and get up before dawn, something they normally hate to do because of the danger of meeting a leopard. Despite their deep-seated fear, the apes would set out on a long trek to a specific fig tree where they had recently eaten. Their goal was to beat the early fig rush by other animals, from squirrels to hornbills. Remarkably, the chimps would get up earlier for trees far from their nests than for those nearby, arriving at about the same time at both. This suggests calculation of travel time based on distance. All this makes Janmaat believe that Taï chimpanzees actively recall previous experiences in order to plan for a plentiful breakfast.[21]

In a classical experiment, Nicky Clayton at Cambridge University studied western scrub jays (a corvid) to see what they remembered about foods they'd cached. The birds were given different items to hide, some perishable (wax worms), others durable (peanuts). Four hours later the jays looked for the worms—their favorite food—before they looked for nuts, but five days later their response was reversed. They didn't even bother to look for the worms, which by that time would have spoiled and become disgusting. But they did remember the peanut locations after this long time. This ingenious study included several controls, allowing Clayton to conclude that jays recall what items they put where and when. The birds had to have manipulated this information inside their heads to make the right choices.[22]

We also have studies of *metacognition*, which refers to the knowing of knowing. Let's say someone asks me if I'd rather answer a question about 1970s pop stars or about science fiction movies. I'd right away pick the first category, because that's what I'm better at. I know what I know. These kinds of experiments have been conducted with animals (monkeys, apes, birds, dolphins, rats), showing that they too have different levels of confidence about what they know. They perform some tasks without hesitation, but at other times they can't make up their mind, exhibiting doubt. In one early study, a dolphin named Natua was asked to discriminate between a high tone and a low tone. His level of confidence was quite manifest. He swam at different speeds toward the response depending on how easy or hard it was to tell the tones apart. When they were quite distinct, Natua swam at full speed with a bow wave that threatened to soak the electronic apparatus. The scientists had to cover it with plastic sheets. If the tones were similar, though, Natua slowed down, waggled his head, and wavered. Instead of touching one of the paddles to make his choice, he selected the opt-out paddle (asking for a new trial), which meant he knew that he'd probably flunk the task. This is metacognition at work, which may involve consciousness as it requires animals to judge the accuracy of their own memory and perception.[23]

Even if this and other studies fail to directly tell us—the way Proust did so eloquently about himself—how aware animals are of their own memories, it is hard to deny the possibility that animals consciously travel along the time dimension and rack their brains for knowledge and experiences.[24] We now have the beginnings of an idea of what consciousness is good for and why it evolved. This argument could be extended to the emotions by arguing that sentience is not enough of a qualification for some animals. Sentience is a general reference to experiencing things, which can be done entirely unconsciously. For species with substantial brains, however, such as all mammals and birds, we need to add consciousness as an option, not just for memories and thinking but also for their emotional lives. I suspect that animals capable of consciously probing their experiences and memories also have the capacity to explicitly recognize the bodily upheavals that we call emotions. It probably helps their decision making to realize how they feel about events in their environment.

All in all, my discussion here distinguishes three levels of sentience.

The first level is sensitivity in a broad sense to the environment and one's own internal state so as to maintain homeostasis and safeguard one's existence. Self-preserving sentience, which may be fully unconscious and automated, characterizes every plant, animal, and other organism and may be the basis of all higher forms. The second level is sentience in the narrow sense, relating to experiencing pleasure, pain, and other sensations to the point that they can be remembered. This form of sentience, which permits learning and the modification of behavior, is best assumed in every animal with a brain, regardless of brain size. And the third level is consciousness, where internal states and external situations are not just remembered but evaluated, judged, and logically connected, such as was done by the hero of Chrysippus's tale. Conscious sentience serves both feelings and problem solving. We don't know when and where it began, but my guess is that it was relatively early in evolution.

Evolution Minus Miracles

In 2016, I was one of the organizers of an international conference about emotions and feelings in humans and animals, in Erice, an ancient Sicilian fortress city atop a 2,500-foot mountain. In between sessions, walking through the winding cobblestone streets with a splendid view of the Mediterranean, Jaak Panksepp and I talked about animals' feelings. I expressed my reluctance to be specific, saying, "I think I know what they feel, but it remains speculation." Jaak, with his kind, melancholic face, shook his head. "First of all, Frans," he replied, "there is solid evidence for animal feelings. Second, what's wrong with a few educated guesses?" He felt I should come out and be more explicit about my impressions. I now believe he was right, and I will try to express his opinion and explain why he had to fight for it all his life.

Panksepp, who sadly passed away about a year after that conference, was of extraordinary importance to affective neuroscience, a field he founded. He placed human and animal emotions on a continuum and was the first to develop a neuroscience covering all of it. He had to resist establishment forces, the most formidable one being B. F. Skinner's school of behaviorism, according to which human emotions are irrelevant and animal emotions suspect. Ridicule was heaped onto him for wanting to study the neuroscience of affect, so Panksepp never received much funding for his work. Despite the lack of money, however, he did more than almost anyone to make animal emotions a respectable topic, and he became known for his studies of joy, play, and laughter in rats based on ultrasonic vocalizations. Rats actively seek out tickling fingers, he found, probably rewarded by opioids in their brains. His work situated the emotions in ancient subcortical brain areas shared across all vertebrates rather than in the recently expanded cerebral cortex. His 1998 magnum opus, *Affective Neuroscience: The Foundations of Human and Animal Emotions*, became a best seller by academic standards. He was ahead of his time and influenced many animal scientists, including Temple Grandin and me.

At the 2016 Erice conference, Panksepp had a long standoff with Lisa Feldman Barrett, who considers emotions to be mental constructs that vary with language and culture. Instead of being hard-wired, in her view, they are woven together from past experiences and moment-to-moment judgments of reality. As a result, it is impossible to clearly pinpoint particular emotions.[25] Her position was almost the exact opposite of Panksepp's subcortical emphasis. Neither scientist was ready to give in, and both kept repeating their arguments the way people do when they aren't listening. I didn't think we needed such a heated confrontation, though, because as soon as one draws a sharp line between emotions and feelings, both positions make sense. Panksepp was mostly talking about the emotions, and Feldman Barrett about feelings. For her, feelings and emotions are one, but for Panksepp, me, and many other scientists, they are to be kept apart. Emotions are observable and measurable, reflected in bodily changes and actions. Since human bodies are the same across the globe, emotions are by and large universal, including what happens

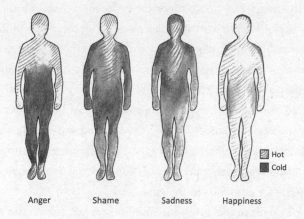

Anger Shame Sadness Happiness

Emotions affect the body as much as the mind. Asked which body parts they sense the most during certain emotions, people from three different cultures colored in silhouette figures. They agreed that anger is felt mostly in the head and torso and happiness all over the body. Shame, by contrast, heats up head and cheeks, but cools off the rest of the body, whereas during sadness most of the body is numb. Based on a study by Nummenmaa et al. (2014).

to us when we fall in love, have fun, or get mad. That is why we never feel emotionally disconnected even in a country where we don't speak the language. Feelings, on the other hand, are private experiences, varying from place to place and from person to person. What one person experiences as pain, another may feel as pleasure. There is no simple one-to-one mapping between emotions and feelings. Every language has its own concepts to describe subjective states, and people bring different backgrounds and experiences to how they feel and why.

The body is heavily implicated, though. While we are describing how we feel, we use visceral language, place our hand over our heart or stomach, ball our fists, clasp our head, or clutch ourselves as if we're falling apart. Crying, for example, is much more than a sound. We have trouble breathing, our heartbeat becomes irregular, our diaphragm is weighed down, we have a lump in our throat, our face becomes a puddle. When we cry, our whole body cries. William James went even further, saying that bodily changes are not so much the expression of an emotion—they *are* the emotion. While this issue keeps being debated, a Finnish team led by Lauri Nummenmaa mapped the body regions involved in certain emotions. They asked subjects to mark body maps to indicate the regions they associate with them. The digestive system and the throat are implied in disgust, the upper limbs in anger and happiness, and the stomach in fear and anxiety. Since the marked areas were strikingly similar for speakers of Finnish, Swedish, and Taiwanese—three largely unrelated languages—the investigators concluded that different cultures must experience emotions in the same way.[26]

This by no means rules out variation in how we discuss our feelings, however. Having French in-laws, I am always struck by how the Dutch convey their feelings in moderate terms, trying to sound calm and reasonable, whereas the French may go all out and grow lyrical and passionate, especially when it comes to love and food. Having been together for many decades, my wife and I are unfazed by these cultural differences even though they still occasionally cause misunderstandings, hilarity, or both. Yet even if the Dutch and the French seem from different planets in how they talk about their sentiments—thus supporting Feldman Barrett's thesis that feelings are constructs—the cultural barrier falls away when it comes to our bodies, voices, and faces. The disappointment of losing Dutch and French soccer fans looks exactly the same.

Much of the confusion boils down to the language filters through which science inspects human emotions. We focus on verbalized experiences while emphasizing linguistic nuances, paying almost more attention to the labels than to the feelings they are supposed to capture. Panksepp's bottom-up neuroscience, on the other hand, starts from so deeply inside the brain that labeling and linguistic concepts are barely relevant. Yet even though feelings were never his focus, Panksepp was convinced that they are always there, not just for humans but also for rats. They are just part of the emotions.

One of the best pieces of evidence is how animals react to drugs that induce pleasurable or euphoric states in humans. How these drugs alter the human brain is well understood. Rats get attracted to exactly the same drugs and undergo the same brain changes. Indeed, the way they respond to a particular new drug (by seeking or avoiding its administration) is a perfect indicator of whether humans will consider the same drug hedonic or aversive. This is hard to explain without assuming shared subjective experiences.

But not everyone is happy with this implication. It is still common to downplay animal feelings or to hide allusions to them in clouds of quotation marks and euphemisms. In 1949, the Swiss physiologist Walter Hess received the Nobel Prize for his discovery that aggressive responses could be induced in cats by electrically stimulating their hypothalamus. A bristling, hissing cat would arch her back, lash her tail, and extend her claws, ready to attack. She'd also show high blood pressure, dilation of the pupils, and other signs of being infuriated. As soon as the stimulation was turned off, however, the cat would calm down and act normally again. Hess spoke only of "sham-rage," however, thus covering up the emotional content of the cat's behavior. After his retirement, he regretted having done so and admitted he had used this evasive term only to avoid the wrath of American investigators, who had trouble imagining that a full-blown emotion could be triggered in a subcortical area of the brain. In fact, Hess said, he had always felt that his cats must have felt genuine rage.[27]

When Europeans, such as Hess or Panksepp, refer to "American investigators" and their qualms, they mean behaviorists. Even though this school of thought has had an international reach, its outlook and radical indoctrination are most keenly felt at North American universities. Behaviorism started out well enough with the goal of developing a unifying framework to explain both human and animal behavior. It got its name from its focus on observable behavior and its contempt for unobservables, such as consciousness, thoughts, and feelings. Behaviorists said psychology needed to cast off the "yoke of consciousness."[28] It needed to talk less, or not at all, about what goes on inside the mind and more about actual behavior.

A major deviation from this commendable agenda occurred half a century later, however, when the cognitive revolution came along. In the 1960s, psychologists began to highlight mental processes in our species and explore consciousness and thinking. They criticized behaviorism as too narrow and pushed it aside. At this time, behaviorism could have renewed itself by adopting a few crucial cognitive concepts and gone with the flow. Instead, it chose to dislodge our species from the rest of the animal kingdom. It was obviously hard for them to argue that humans can't think or are unaware of themselves but for animals, this remained a defensible position. Behaviorism doubled down on animals as stimulus-response machines, while applying its approach more lightly and selectively to humans. In doing so, behaviorism opened a gap between humans and all other species, which only widened with time.

As a result, psychology departments all over the world began to house two quite different kinds of faculty. Those who worked on human behavior blithely assumed a wide range of complex mental processes accompanied by a high degree of awareness. The capacities they proposed could be truly convoluted, such as one person knowing that another one knows that the first knows something that they don't know. In contrast, the faculty occupied with animals, known as comparative psychologists, adopted the exact opposite approach, studiously avoiding any mention of mental processes and preferring the simplest possible account. They explained animal behavior in terms of learning from experience, irrespective of whether the animal had a large or small brain, was predator or prey, flew or swam, was cold- or warm-blooded, and so on. Scientists who dared speculate about special abilities related to a species's natural history could count on fierce resistance, given that exceptions to the "law of effect" weren't welcome. Since biology, ecology, and evolution were kept out of behaviorism, one wonders how it survived for as long as it did.

The dichotomy between human psychologists, making incredibly generous assumptions, and animal psychologists, being excessively stingy, created a problem that William James saw coming long ago. He stressed continuity between humans and other animals:

> *The demand for continuity has proved itself to possess true prophetic power. We ought therefore ourselves sincerely to try every possible mode of conceiving the dawn of consciousness so that it may not appear equivalent to the irruption into the universe of a new nature, non-existent until then.*[29]

Unfortunately, an "irruption" (a violent, forcible entry) was the only way to reconcile the contrasting perceptions of human and animal intelligence. This is why we so often hear that an extraordinary jump must have taken place during human evolution. Obviously, no modern scholar would dare speak of a Divine Spark, let alone special creation, but the idea is familiar enough. The current endless stream of books on what makes humans different reveals, as the promotion for one book put it, "humankind in its glorious uniqueness—one foot planted firmly among all of the creatures we've evolved alongside, and the other in the special place of self-awareness and understanding that we alone occupy in the universe."[30] Every book on human exceptionalism tells a slightly different story about how we got so lucky: through a special (but always mysterious) brain process, the impact of culture and civilization, or an accumulation of small changes with huge consequences. Friedrich Engels, the German philosopher friend of Karl Marx, even gave us an essay on "The Part Played by Labour in the Transition from Ape to Man."

Regardless of the theory, however, we're asked to believe in a pretzel-like twist rather than the usual slow and smooth course of evolution. But these twists are needed only because science has neglected what animals are capable of. We have for the longest time harbored such minimalist assumptions about animals that, in comparison, our own cognitive achievements seem totally out of reach. But what if animal intelligence is nowhere near rock bottom?

Today we are in the midst of a belated cognitive revolution with regard to our fellow species. A younger generation of scientists has abandoned the taboos that held us back for so long. The level at which we put other species is inching up day by day. The Internet regularly features exciting scientific breakthroughs in *evolutionary cognition*—the study of human and animal intelligence from an evolutionary perspective— accompanied by striking videos of apes, corvids, dolphins, elephants, and so on, demonstrating causal reasoning, theory-of-mind, planning, self-awareness, and cultural transmission. This new research has increased the regard in which we hold animal intelligence so dramatically that we no longer need miracles to explain the human mind. Its basic features have been around for ages.

In the meantime, neuroscience is breaking open the black box to look inside the brain, offering accounts of how animals solve problems that rely less and less on the learning theories of the past. Behaviorism is dying a slow death, lifting its head only now and then to feebly try putting a brake on these developments. It is this brake that Panksepp experienced all his life, when behaviorism was still in full swing. The prevailing robotic view of animals was anathema to him, while he complained about the "terminal agnosticism" that prevented anyone from taking a stance on the origins of consciousness.

In the West, we have always been enamored with mechanistic metaphors. Biological processes that we have trouble understanding are compared with machines. We do understand machines, because we design them ourselves. We say that the heart is like a pump, the body like an automaton, and the brain like a computer. Finding biology too murky and messy to be wholeheartedly embraced, we try turning it into something like Newtonian physics. Most famously, the seventeenth-century French philosopher René Descartes tried to fold the passions into this mechanical view:

I should like you to consider that these functions (including passion, appe-
tite, memory, and imagination) follow from the mere arrangement of the
machine's organs every bit as naturally as the movements of a clock or other
automaton follow from the arrangement of its counter-weights and wheels.[31]

The clock metaphor has been debated for nearly as long as it has been around; its obvious flaw is that in biology everything grows and develops together and is fully connected with everything. The brain looks much more like gelatinous soup than a machine, and it has billions and billions of connections, incredibly integrated at every level. Moreover it is part of the body and should never be considered in isolation. By contrast, man-made contraptions are put together from discrete parts that are produced separately and meet for the first time on the watchmaker's desk. After they have been put together, they neither converse with nor depend on each other, except for a few preassigned connections. There is no distant communication, while inside the human body we discover new instances all the time, such as the connection between the gut microbiome and the brain, or the cardiac synchronization between mother and fetus.[32] In a clock, each part remains more or less independent, so the clock can be taken apart and put back together into a functioning whole. But no organism would permit such rough treatment. You remove one part, let's say the liver, and you can forget about the whole. Your "machine" is now broken. In fact, it is not broken but dead!

Panksepp had little patience with the notion that animals are best understood as input-output systems with a limited set of responses. Organisms have nothing in common with machines, and all those clock and computer metaphors are distinctly unhelpful. Instead, he showed a genuine interest in the inner lives of animals and, like any biologist, assumed their continuity with humans.

The fact that we cannot directly detect what animals feel is hardly an obstacle. After all, science has a long tradition of working with unobservables. Evolution by natural selection is not directly visible, nor is continental drift or the big bang, yet all these theories are so well supported that we treat them almost as facts. Or take a staple of psychology, theory-of-mind. No one has ever seen how it works, yet it is considered a milestone of child development. In all these cases, we gather evidence and see how it fits our theories. Even the notion that the earth is a sphere lacked direct confirmation until 1967, when the first color picture of our planet was taken from outer space. This is why we should never accept the common argument that animal feelings and consciousness are off limits to science because we can't see them. Panksepp wisely urged:

> *If we are going to entertain the existence of experiential states, such as consciousness, in other animals, we must be willing to work at a theoretical level where arguments are adjudicated by the weight of evidence rather than definitive proof.*[33]

No Fish No Cry

Strange as it may sound coming from a primatologist, I am a great fish fan. As a boy, I would go out on my bike on Saturdays to catch stickleback fish, salamanders, and all sorts of aquatic life. I kept these animals in a growing array of pots and buckets until I received my first aquarium as a birthday present. I have never lived without an aquarium since, and I now have two large planted freshwater tanks built into the walls of my home.

Fish in my hands almost never die anymore. I have one large Plecostomus, who must be over twenty-five years old, and a small group of clown loaches aged at least fifteen. Even though the loaches look a bit like Nemo—the clownfish of animated movie fame—it's only the bold patterning that they have in common: clownfish live in the ocean, whereas clown loaches are inland water fish of a different family. They are big and plump yet agile swimmers, fun to watch when they tumble

Fish can be quite sociable, not just in large schools but also in small schools in which they recognize each other individually. Clown loaches are tropical freshwater fish that often swim, tumble, and hang out together.

around as a group. Always hanging out together, they make lots of body contact, often squeezing together into small crevices. The secret of a good tank is to have plenty of hiding places, and as soon as a loach sees one of his friends in one, he or she joins them, whereupon the two sit tightly packed together looking out. Often all six of them pile on top of each other. I say "friends" because they recognize each other. I learned this the hard way when on a few occasions I tried to introduce new loaches. The newcomers never encountered any aggression, the way territorial fish chase others off, but they got the cold shoulder to the point that they never integrated with the established clique.

I enjoy the sociability of my loaches and all the other fish interactions that are far more complex than most people realize. Some mated pairs get along fine and swim smoothly side by side wherever they go, whereas other pairs are always squabbling and posturing and barely let each other eat. With such poor bonding, I know they'll never breed. Some fish watch over their young: many cichlids do, as did the sticklebacks that I kept as a child. After fertilization, father stickleback fans the eggs to provide them with extra oxygen, and he keeps the fry together for a few days after hatching, sucking up straying babies only to spit them back into the nest. To watch these interactions up close is the privilege of the aquarist and also the reason I never understand the low regard in which fish are held. It's as if they were some lesser life-form, not worthy of the concerns we have for other animals.

Naturally, discussions about sentience often turn to the question of whether fish feel pain, as they have for fifty years. No one will kick a pregnant dog and interpret her whimpers as the rattling of gears, the way a prominent Cartesian allegedly did (whereas Descartes himself doted on his dog), but when it comes to fish, people do harbor doubts. Part of the confusion is that fish don't necessarily feel pain when they flop around or escape from danger. Like many animals, they have receptors on neuron axons that react to peripheral tissue damage. This is known as *nociception*, which is automatic, the way we pull back our finger upon touching a hot stove, even before we've become aware of the pain. Nociceptors send signals to the brain, which instructs the body to get rid of or move away from the threat. It has long been argued that fish have only this reflex-like pain system.

Does this mean that a fish wriggling on a hook feels nothing? The fishing industry certainly would like us to think so. Many studies argue that since fish lack a cerebral cortex, they aren't wired for pain sensation. Additional confusion is caused by the fact that they don't utter distress calls. Because we humans take high-pitched noise as the best indicator of pain, we wonder how bad it can be for fish if they never cry. Fish have other methods of communication, though. A good illustration took place among the goldfish in one of my backyard ponds.

Although I love all wildlife, I must say that I draw the line at herons. These beautiful birds are adapted for spearing prey, but they do their job a bit too well. Goldfish, by contrast, are bred to stand out, which is a hugely maladaptive trait. As a result, a heron can empty out a goldfish pond in a few hours. One day after I saw a heron close to the edge of one of my ponds, I decided to put up a netted barrier. No longer able to hunt there, the heron did not return.

But one of the goldfish got caught in the net, which hung partly in the water. I cut it loose, but before then this fish must have struggled for quite a while—it was marked with a white band all around where the net had dislodged its golden scales. Afterward all the other fish were unusually timid, and for days they would not come out of hiding, not even for food. They might have noticed the struggle of their companion, which probably had gone on for hours. But interestingly, the fish in my second pond, which is separate from the first, were equally terrified. They, too, stayed in the deep. Given that I don't believe in telepathy, this made no sense. They could have had no direct knowledge of the fish's struggle.

Almost a century ago an Austrian scientist discovered *Schreckstoff*. The German verb *schrecken* refers to our reaction when something suddenly scares us. If a bear were to enter through my window, for example, I would "schreck" myself to death. *Stoff* means "matter." So *Schreckstoff* is a substance that carries the chemical message of a frightened sender who has most likely been physically damaged or killed by a predator. Whereas it is too late for the sender, the release of *Schreckstoff* warns all other fish and gives them time to take countermeasures so the same thing doesn't happen to them. That the alarm signal only benefits its recipients, and not the sender itself, is puzzling enough, but for me the question was how the stuff might have jumped between two bodies of water. Only when I realized that a single filter cleans both my ponds did I understand.

It took my hurt goldfish two months to recover (its white band fading), while the other fish returned to normal within a week. Without having any direct knowledge of the traumatic event, they had shown the right antipredator responses as a result of this chemical warning system. But even though science now knows the active ingredient (a sugar-like molecule) of *Schreckstoff*, this doesn't resolve the issue of what fish feel.[34]

Physiologically, fish are remarkably similar to mammals. They have an adrenaline response to sudden events, and high cortisol levels when crowded or harassed. A fish who hides in the farthest corner of the tank all day because of an intolerant territory owner may literally die of stress. Fish also have dopamine, serotonin, and isotocin. The latter is their equivalent of oxytocin, which plays a role in social behavior.

We shouldn't be surprised therefore by studies of depression in fish. One induced it by getting zebrafish addicted to ethanol. After weeks of partying, the experimenters cut off the ethanol supply, forcing the fish into withdrawal. Like depressed people, they lost interest in life and became passive and withdrawn. Rather than racing along the water surface as they are wont to do, they dropped to the bottom of the tank, where they stayed more or less motionless. Fish are normally curious and fare best in enriched environments, but now they were bored and didn't even explore their tank. Mind you, to speak of boredom or depression in fish is not just human projection, because if these miserable-looking fish are given an antidepressant, such as diazepam, they liven up and spend more time near the top of the tank. That the same drug helps both fish and humans hints at profound neurological similarities.[35]

The story of pain is similar. Victoria Braithwaite, a British fisheries scientist, offers examples of fish intelligence in her 2010 book *Do Fish Feel Pain?* and also shows how fish respond to negative stimuli. When they are injected with irritating chemicals under their skin, such as vinegar, they rub themselves in the gravel of the tank to get rid of it. They lose their appetite and are too distracted to avoid novel objects. When they are given a pain-killer such as morphine, however, these reactions disappear. Fish also try to avoid pain, and not just in the reflex-like manner you'd expect from a nociception system. They remember where they encountered painful stimulation and avoid these places. Using the same argument that we applied to crabs, the idea here is that for negative stimuli to be remembered, they must have been felt. As a result of this and other studies, the consensus view is now that fish do feel pain.[36]

Readers may well ask why it has taken so long to reach this conclusion, but a parallel case is even more baffling. For the longest time, science felt the same about human babies. Infants were considered sub-human organisms that produced "random sounds," smiled simply as a result of "gas," and couldn't feel pain. Serious scientists conducted torturous experiments on human infants with needle pricks, hot and cold water, and head restraints, to make the point that they feel nothing. The babies' reactions were considered emotion-free reflexes. As a result, doctors routinely hurt infants (such as during circumcision or invasive surgery) without the benefit of pain-killing anesthesia. They only gave them curare, a muscle relaxant, which conveniently kept the infants from resisting what was being done to them. Only in the 1980s did medical procedures change, when it was revealed that babies have a full-blown pain response with grimacing and crying. Today we read about these experiments with disbelief. One wonders if their pain response couldn't have been noticed earlier![37]

Scientific skepticism about pain applies not just to animals, therefore, but to any organism that fails to talk. It is as if science pays attention to feelings only if they come with an explicit verbal statement, such as "I felt a sharp pain when you did that!" The importance we attach to language is just ridiculous. It has given us more than a century of agnosticism with regard to wordless pain and consciousness.

Transparency

Research into animal intelligence and emotions has had the paradoxical effect of yielding arguments against research itself. My own findings are sometimes angrily thrown back at me. Should we inject fish with vinegar, subject monkeys to cognitive tasks, keep dolphins in captivity, or even have pets at home? Some argue that behavioral research is unnecessary, because *of course* animals are smart and have human-like emotions. *Everyone* knows this! I beg to differ—if it were true, we wouldn't have had to fight so hard to get these ideas accepted. Let's not forget that for ages animals were depicted as dumb automatons without meaningful sensations and emotions. The "everyone knows" argument doesn't cut it.

If humans had permanently kept their distance from animals, and never mingled with them or explored their abilities, we'd know next to nothing about them and probably wouldn't care. We rarely worry about anything that doesn't touch us. I'm a firm believer, therefore, that the fact that many people have animal companions at home and regularly visit zoos and nature reserves, where they see animals up close, has a huge positive impact on our relations with our fellow species. Many city dwellers, who are more and more estranged from nature, hold a Disney-fied view that doesn't correspond with the harsh realities of survival. Being with animals profoundly shapes our perceptions and nudges us to learn more about them and care about conserving them. Watching entire school classes of children running around at the zoo filling in their teachers' questionnaires makes me optimistic, because I see enthu-siasm and a hunger for knowledge. It all boils down to that which evolu-tionary biologist E. O. Wilson described as *biophilia*: our instinctive bond with nature and other animals. We have a long history of close interac-tion with animals, both for pleasure and for subsistence, the abandon-ment of which would not necessarily be a good thing for them or for us. It would leave animals even more out in the cold than they already are.

Things might be different if we still had pristine habitats available for animals to disappear into, but unfortunately, this is not the world we live in anymore. The whole idea of liberty for animals is in question. Some domesticated species that sought contact with us long ago now depend on us. Wild animals, too, often have no other option than to live close to us or at least under our protection. Many have been forced to carve out a survival niche in our expanding cities. A great deal of animal evolution has shifted to human-made environments, such as the urban coyotes in North America (I have them in my backyard) and the ring-necked parakeets—colorful tropical birds—now screeching around by the thousands in European cities. City animals actually change their gene pools while adapting to their new environments.[38] On the other hand, animals that have lost their original habitat and are unable to adjust to ours are in deep trouble. I could give many examples, but sadly our clos-est relatives, the apes, are first among them.

To put it in the starkest possible terms: if I were born tomorrow as an orangutan, and you were to offer me a choice between living in the jungles of Borneo or at one of the world's finest zoos, I would probably not choose Borneo. Depressing pictures reach us of juvenile orangutans clinging to the last little tree that remains of their burned-down forest. When displaced orangutans try to eat fruits grown by farmers, they are treated as pests and shot. Others end up in an overflowing Indonesian sanctuary. This large ape species requires high-quality food and cannot simply be relocated to other habitats, most of which are degraded and shrinking. Obviously, orangutan rehabilitation centers need all the support we can offer them, but we have a serious ape "refugee" crisis on our hands with little hope for a solution. An estimated one hundred thousand orangutans (about half the total population!) have disappeared from Borneo in the last two decades. Similar problems affect other critically endangered species, such as rhinos (which travel the plains of Kenya with armed bodyguards), mountain gorillas (less than one thousand are left in the wild), California condors (brought back from the brink by a captive breeding program), vaquitas (less than thirty of these small porpoises still live in the Sea of Cortez), and the list goes on. We can keep idealizing the natural habitat as the only place where wild animals belong and are free, but what is freedom without survival?

With respect to animals in research, the landscape is changing. The more an animal is like us, the easier it is to extend our moral outlook to it, which is why chimpanzees were the first to benefit from the attitude changes currently under way. In 2000 New Zealand passed legislation banning research on great apes, while Spain adopted a resolution to grant these animals legal rights. Neither country had been doing actual ape research, however. As a result, I could not resist remarking to a Spanish journalist that I would have been more impressed if they'd abolished bullfighting. Only when the Netherlands and Japan passed similar laws did the movement to improve the status of apes begin to make a difference, because both countries outlawed what they had been practicing. Ruling out euthanasia as a means of population control, both governments faced the expensive need to find a home for ex-laboratory chimpanzees, some of which required special precautions and care, as they had been infected with diseases. In 2013, the United States joined the club, not by outlawing the use of apes in biomedical research but by cutting off funding, which amounts to the same thing.

I am in full support of this decision, even though it has also curbed the kind of noninvasive behavioral studies that I conduct myself. I'm a longtime board member of ChimpHaven, in Louisiana, which is the world's largest retirement facility for chimpanzees. The apes arrive from laboratories and facilities all over the country, to be released onto large forested islands, where they live out their lives. ChimpHaven offers the best environment one can imagine outside the natural habitat. Given the current demand, we are busy building new islands.

For the remaining animals in research and the agricultural industry, I place my hopes on transparency. It's up to society to decide what kind of relations we'll have with animals, and what kind of uses we'll permit, but it is absolutely essential to bring animals out of the shadows. We barely know what's going on at many places, which makes it easy to act as if nothing is the matter. We need research facilities with open-door policies and farms with an obligation to show how they keep their animals. Ideally, meat packages in the supermarket would feature a scan bar that allows us to call up pictures (taken by an independent agency) on our smartphones so that we can judge the animal's living conditions for ourselves. If we made all locations with captive animals as public as zoos, matters would improve in a hurry. Public pressure and consumer preferences would do the job.

Having worked in primate facilities for years, I think the biggest step forward there would be a law that says we can't keep any primates unless we house all of them socially. Still too many facilities have batteries of single-caged macaques. Whatever research we deem essential, the very least we should offer these animals is a social life. Admittedly, such a life is not without stress—it is in fact full of drama and fights.

But it also offers bonding, grooming, and play. Since I've always worked with group-living primates, I know from experience that they thrive in a social setting. Whether they bicker or groom, they are made to live together. One illustration of how important this is to them: one day at the Yerkes Field Station, we gave our chimps a brand-new climbing structure in their outdoor area, a huge wooden frame with ropes and nests high above the ground, from which they would be able to scan for miles. We confined the whole colony indoors for a few weeks while we worked hard to erect it outside. We were so proud of our design, and so looking forward to the apes' reaction, that when we released the colony, we expected they'd rush into the structure and enjoy the sights. But during their time indoors, they had been separated from one another in various areas, and now they clearly had a different idea.

The first thing that happened was a gigantic emotional reunion. They barely glanced at the new construction, having eyes only for each other while hooting with excitement and joy. They walked from one to the other, touching, kissing, and embracing their long-lost friends and family. For them, the big moment was intensely social. Inspection of the new climbing frame could wait. It taught me once again that whenever we seek to provide optimal housing, social life beats physical conditions anytime.

A common argument by researchers against group housing is that certain procedures require them to have daily access to the animals. This is a poor argument, however, given how easy it is to train primates to come out of their group. All you need to do is call them by name and open a door. In fact, many experiments could be done on a volunteer basis provided they are enjoyable for the animals. At the Primate Research Institute in Japan, the outdoor enclosure has cubicles that the chimpanzees can enter any time to work by themselves on a computer screen. They can also leave any time they want. A video camera tells the investigators whose digital data they are looking at. We really don't need constant access anymore, given that advances in wireless technology and microchips allow primates to be semi-free while being studied. At the Yerkes Field Station, for example, rhesus macaques live in large outdoor corrals in groups of about one hundred each. With a little creativity and technical know-how, these conditions can accommodate almost any kind of research. Ideally, primate facilities would ditch all their small cages and restraining chairs and monitor their animals' vital functions while they enjoy the company of others. It would be better for the monkeys and would produce better science. In many places, scientists and IT specialists are joining forces to achieve this goal. To force facilities into this direction, transparency will be key. Primate centers should open their doors to the press and public, allowing them to come take a look, either in reality or via live webcams. As social primates ourselves, most people understand at an intuitive level what living conditions best suit monkeys.

And so we have drifted here from a discussion of sentience to one on how we ought to treat the animals in our care. This transition is natural and timely, now that both science and society are ready to abandon the mechanistic view of animals. As long as we took this view, no one needed to worry about ethics, which sadly may have been part of its appeal. If, on the other hand, animals are sentient beings, we have an obligation to take their situation and suffering into account. This is where we are now. We behavioral scientists urgently need to get involved, not only because we are users of animals, which would be reason enough, but also because we are at the forefront of the changing perceptions of animal intelligence and emotions. We are pushing for a new appreciation of animals, so we'd better help implement the needed changes. We have instruments by which to know specifically which conditions are beneficial or harmful for animals. We can offer animals a choice of different environments to see which ones they prefer: Will chickens seek out a hard surface or dirt? Do pigs really like mud? Animal well-being is measurable, and its study is becoming a science in and of itself, which of course would never have happened if we were still convinced that animals feel nothing.

8 | CONCLUSION

Early ethologists studied fish, birds, and rodents to find out how behavior patterns hang together. If they occurred in sequence, such as freezing and escape, or threat and attack, we reasoned, they probably shared a motivation. I was a student at the time, and we talked about little else than these clusters of behavior, known as *behavior systems,* presented in elaborate diagrams to illustrate how animals prioritize them. Animals moved between behaviors within each system, we observed, the way the male stickleback performed his zig-zag dance, to achieve a goal, such as a female dropping off her eggs in his nest. The system approach was elegant and objective, but something was missing: Where did those underlying motivations come from? What were they? As we discussed this question, we always carefully avoided any allusion to the emotions. But in retrospect, the source of many behavior systems looked suspiciously like internal states such as fear and anger.

The silence surrounding the emotions was all the more puzzling if we consider leading alternative proposals for the motivation of animal behavior. The prevailing view was that animals had instincts, a series of inborn actions triggered by a particular situation, or preprogrammed simple responses, such as one kind of action adapted to one kind of context. This sounds cumbersome, though, because it could lead only to rigid behavior, which would be a disaster under changing circumstances.

Imagine that a male animal, like a machine, was programmed to react to the sight of a female with automatic sexual arousal, courtship, approach, and mounting. Sometimes this might work, but what if the object of attention was fiercely unwilling? What if a jealous dominant male sat nearby? Or imagine a predator coming around the corner at the wrong moment. Clearly, a fully automated response might get our male in trouble. The reason scientists rarely talk about instincts anymore is that they are just too inflexible.

Instead, thinking in terms of emotions, we consider that the sight of an attractive mate induces a strong desire together with a careful evaluation of the circumstances. The desire will urge the individual to strive for the best possible outcome. Other emotions do the same, such as when animals deal with predators, seek to protect their young, strive for a better place in the hierarchy, are interested in the same food as others, and so on. All these situations arouse emotions, which generally have the organism's best interest at heart. But they merely prime body and mind. They don't dictate any specific course of action. Sometimes freezing is better than fleeing, sometimes sharing food yields more than fighting, and sometimes a sex partner will need to be led to a hidden location before copulation can take place. Emotions allow this flexibility.

The field of artificial intelligence recognizes this advantage; hence its attempts to equip robots with "emotions." This is done partly to facilitate interactions with humans but also to provide a logical architecture for robot behavior. Emotions have the benefit that they direct attention, make events memorable, and prepare for engagement with the environment. They are a better way to structure behavior than giving machines piecemeal instructions for every situation they will encounter. When scientists program emotion-based robots, they come up with interesting definitions, such as: "the robot is *happy* if there is nothing wrong with the present situation. It will be particularly happy if it has been using its motors a lot or is in the process of getting new energy."[1] That robot emotions are a growing field, known as *affective computing,* suggests that equipping entities with action-oriented internal states is the best way to organize behavior, just as evolution has done for us. It is how we operate, and how most animals operate. We are emotional beings through and through.[2]

For me, the question has never been whether animals have emotions, but how science could have overlooked them for so long. It didn't do so originally—remember Darwin's pioneering book—but it certainly has done so recently. Why did we go out of our way to deny or deride something so obvious? The reason, of course, is that we associate emotions with feelings, a notoriously tricky topic even in our species. Feelings happen when emotions bubble to the surface so that we become aware of them. When we are conscious of our emotions, we are able to express them in words and make others aware of them: they see emotions in our face, but they get the feelings from our mouth. We say we are "happy," and people believe us, unless of course they can see for themselves that we're not. Sometimes a human couple acts as if they were happy but then divorce a month later. People close to the pair probably knew. If not, they will wonder how they could have missed the signs. We are very good at separating reported feelings from visible emotions, and we generally trust the latter better than the former.

The possibility that animals experience emotions the way we do makes many hard-nosed scientists feel queasy, partly because animals never report any feelings, and partly because the existence of feelings presupposes a level of consciousness that these scientists are unwilling to grant to animals. But considering how much animals act like us, share our physiological reactions, have the same facial expressions, and possess the same sort of brains, wouldn't it be strange indeed if their internal experiences were radically different? Language is irrelevant to this question, and the size of our cerebral cortex is hardly a reason to propose a difference. Neuroscience has long ago abandoned the idea that feelings arise there. They come from much deeper inside the brain, the parts closely connected to our bodies. It is even possible that feelings, instead of being a fancy by-product, are an essential part of the emotions. The two may be inseparable. After all, organisms need to sort out which emotions to follow and which ones to suppress or ignore. If becoming aware of one's own emotions is the best way to manage them, then feelings are part and parcel of the emotions, not just for us but for all organisms.

But all right, for the moment all this remains speculation. Feelings are clearly less accessible to science than emotions. One day we may be able to measure the private experiences of other species, but for the moment we have to content ourselves with what is visible on the outside. In this regard, we are beginning to make progress, and I predict that a science of the emotions will be the next frontier in the study of animal behavior. While we are well under way in discovering all sorts of new cognitive capacities, we need to ask what is cognition without the emotions? Emotions infuse everything with meaning and are the main inspiration of cognition, also in our lives. Instead of tiptoeing around them, it's time for us to squarely face the degree to which all animals are driven by them.

ACKNOWLEDGMENTS

Given the centrality of the social domain to my interests as a primatologist, emotions have always been present in the background. They are an undeniable part of primate politics, conflict resolution, bonding, sense of fairness, and cooperation. I started out as an observer of spontaneous social behavior but ended up testing mental capacities, such as face recognition and empathy for another's situation. It is high time, therefore, that I dove more explicitly into the emotions. Hence this book. I look at *Mama's Last Hug* as a companion to my previous book, *Are We Smart Enough to Know How Smart Animals Are?*, which is all about animal intelligence. Even though these two books treat emotions and cognition separately, in real life they are fully integrated.

Luckily, I was trained by an expert in primate facial expressions, Jan van Hooff, at the University of Utrecht. The face being the window to the soul, it is impossible to discuss facial expressions without talking about the emotions. In humans, too, emotion research began with the face. As a result, I grew comfortable with the topic of animal emotions early on, at a time when most scientists still tried to steer clear of it.

I am grateful to the many people who accompanied me on this journey, from colleagues and collaborators to students and postdocs. Just to thank those of the last few years: Sarah Brosnan, Sarah Calcutt, Matthew Campbell, Devyn Carter, Zanna Clay, Tim Eppley, Katie Hall, Victoria Horner, Lisa Parr, Joshua Plotnik, Stephanie Preston, Darby Proctor, Teresa Romero, Malini Suchak, Julia Watzek, and Christine Webb. For help with sections of the book, I thank Victoria Braithwaite, Jan van Hooff, Harry Kunneman, Desmond Morris, and Christine Webb. I appreciate Burgers Zoo, the Yerkes National Primate Research Center, and the Lola ya Bonobo Sanctuary near Kinshasa for opportunities to conduct research, and Emory University and the University of Utrecht for the academic environment and infrastructure to make this kind of work possible. I think fondly of the many monkeys and apes who have participated in and enriched my life, most of all of course Mama, the late matriarch central to this book, who made such a deep impression on me.

I am indebted to my agent Michelle Tessler and my editor at Norton, John Glusman, for their enthusiasm and critical reading of the manuscript. Catherine, my wife, is always there to support and spoil me and to help me stylistically with my daily writing. Nothing feels better than our mutual love and friendship. As a bonus, I have enjoyed many first-hand lessons about human emotions.

NOTES

PROLOGUE

1. B. F. Skinner (1953), p. 160

CHAPTER 1: MAMA'S LAST HUG

1. Mama hugs Jan van Hooff: www.youtube.com/watch?v=INa-oOAexno.
2. Donald O. Hebb (1946), p. 88.
3. Tetsuro Matsuzawa (2011).
4. Otto Adang (1999), p. 116.
5. Bruce Springsteen (2016), p. 78.
6. Robert Yerkes (1941).
7. Steffen Foerster et al. (2016).
8. Barbara King (2013).
9. James Anderson et al. (2010), Dora Biro et al. (2010).
10. Inna Schneiderman et al. (2012), Dirk Scheele et al. (2012).
11. Larry Young & Brian Alexander (2012), Oliver Bosch et al. (2009).
12. Patricia McConnell (2005), p. 253.
13. Geza Teleki (1973).

CHAPTER 2: WINDOW TO THE SOUL

1. Gregory Berns et al. (2013).
2. Paul Ekman (1998), p. 373.
3. Paul Ekman and Wallace Friesen (1971).

4. Irenäus Eibl-Eibesfeldt (1973).
5. Charles Darwin (1872), p. 219.
6. John van Wyhe and Peter Kjærgaard (2015), p. 56.
7. Charles Darwin (1872), p. 142.
8. Kathryn Finlayson et al. (2016), Dale Langford et al. (2010).
9. Matthijs Schilder et al. (1984), Jen Wathan et al. (2015), Mathilde Stomp et al. (2018).
10. Juliane Kaminski et al. (2017).
11. Jan van Hooff (1972).
12. Richard Andrew (1963).
13. Michael Kraus and Teh-Way Chen (2013).
14. David McFarland (1987), p. 151.
15. Anne Burrows et al. (2006).
16. John Lahr (2000), p. 206.
17. Robert Provine (2000).
18. Marina Davila Ross, Susanne Menzler, and Elke Zimmermann (2007).
19. Raoul Schwing et al. (2017).
20. Nadia Ladygina-Kohts (1935).
21. Marc Bekoff (1972).
22. Jessica Flack, Lisa Jeannotte, and Frans de Waal (2004).
23. Richard Alexander (1986).
24. Jaak Panksepp and Jeff Burgdorf (2003), p. 535.
25. Mother chimp tricks her son: www.youtube.com/watch?v=jealP0egJ9k.
26. Lisa Parr et al. (2005).

CHAPTER 3: BODY TO BODY

1. Marian Breland Bailey (1986), p. 107.
2. Nikolaus Troje (2002), for a video: www.biomotionlab.ca/Demos/BMLwalker.html.
3. Frans de Waal and Jennifer Pokorny (2008).
4. *The Correspondence of Charles Darwin*, vol. 2: *1837-1843*.
5. Blaise Pascal, *Pensées* (1669): *"Le cœur a ses raisons, que la raison ne connaît point."*
6. Frans de Waal (2011), p. 194.
7. Daniel Vianna and Pascal Carrive (2005).
8. Lisa Parr (2001).
9. Sarah Calcutt et al. (2017).

10. Ulf Dimberg et al. (2000, 2011).
11. Simon Baron-Cohen (2005), p. 170.
12. David Neal and Tanya Chartrand (2011).
13. Leo Tolstoy (1904), p. 1.
14. Cat Hobaiter and Richard Byrne (2010).
15. Katy Payne (1998), p. 63.
16. Yasuo Nagasaka et al. (2013).
17. Susan Perry et al. (2003).
18. Annika Paukner et al. (2009).
19. Matthew Campbell and Frans de Waal (2011).
20. Ivan Norscia and Elisabetta Palagi (2011).
21. Jeffrey Mogil (2015).
22. Michael Ghiselin (1974), p. 247.
23. Alan Sanfey et al. (2003).
24. Nadia Ladygina-Kohts (1935), p. 121.
25. Carolyn Zahn-Waxler and Marian Radke-Yarrow (1990).
26. Deborah Custance and Jennifer Mayer (2012).
27. Teresa Romero, Miguel Castellanos, and Frans de Waal (2010).
28. Joshua Plotnik and Frans de Waal (2014).
29. Marie Rosenkrantz Lindegaard et al. (2017).
30. Sarah Blaffer Hrdy (2009).
31. Patricia Churchland (2011).
32. Simon Baron-Cohen (2005).
33. Adam Smith (1759), p. 9.
34. James Burkett et al. (2016).
35. Tony Buchanan et al. (2012).
36. Lauren Wispé (1991).
37. Konrad Lorenz (1980).
38. Sarah Brosnan and Frans de Waal (2003a).
39. Koichiro Zamma (2002), p. 11.
40. Tania Singer et al. (2006).
41. Brian Hare and Suzy Kwetuenda (2010), Jinghzi Tan and Hare (2013).
42. Victoria Horner et al. (2011).
43. Ernst Fehr et al. (2018).
44. Shinya Yamamoto, Tatyana Humle, and Masayuki Tanaka (2012).
45. Roger Fouts (1997).
46. Martin Hoffman (1981), p. 133.
47. Inbal Ben-Ami Bartal, Jean Decety, and Peggy Mason (2011).

48. Nobuya Sato et al. (2015).
49. Inbal Ben-Ami Bartal et al. (2016).
50. Felix Warneken and Michael Tomasello (2014).

CHAPTER 4: EMOTIONS THAT MAKE US HUMAN

1. Kerstin Limbrecht-Ecklundt et al. (2013).
2. Joseph LeDoux (2014).
3. Disa Sauter, Oliver Le Guen, and Daniel Haun (2011).
4. Marcel Proust (1982), p. 425.
5. Frans de Waal (1997a).
6. James Henry Leuba (1928), p. 102.
7. Ibid.; http://news.janegoodall.org/2017/11/21/tchimpounga-chimpanzee-of-the-month-wounda/.
8. Edvard Westermarck (1908), p. 38.
9. Malini Suchak and Frans de Waal (2012).
10. Frans de Waal and Lesleigh Luttrell (1988).
11. David Chester and Nathan DeWall (2017).
12. Filippo Aureli et al. (1992).
13. Julia Sliwa and Winrich Freiwald (2017).
14. Mathias Osvath and Helena Osvath (2008).
15. Otto Tinklepaugh (1928).
16. Adam Smith (1776), Chapter II, p. 14.
17. Kimberley Hockings et al. (2007).
18. Fany Brotcorne et al. (2017).
19. Michael Mendl, Oliver Burman, and Elizabeth Paul (2010).
20. Catherine Douglas et al. (2012).
21. Caitlin O'Connell (2015), p. 3.
22. Jessica Tracy (2016); Tracy and David Matsumoto (2008).
23. Jessica Tracy (2016), p. 91.
24. Paul Chen, Roman Carrasco, and Peter Ng (2017).
25. Abraham Maslow (1936).
26. Daniel Fessler (2004).
27. Denver, the guilty dog: www.youtube.com/watch?v=B8ISzf2pryI.
28. Alexandra Horowitz (2009).
29. Konrad Lorenz (1960).
30. Christopher Coe and Leonard Rosenblum (1984), p. 51.

31. From *Chimpanzee Politics* (de Waal, 1982), p. 92.
32. June Tangney and Ronda Dearing (2002), Petra Michl et al. (2014).
33. Winthrop and Luella Kellogg (1933), p. 171.
34. Fausto Caruana et al. (2011).
35. Paul Rozin, Jonathan Haidt, and Clark McCauley (2000); Joshua Tybur, Debra Lieberman, and Vladas Griskevicius (2009).
36. Victoria Horner and Frans de Waal (2009).
37. Erica van de Waal, Christèle Borgeaud, and Andrew Whiten (2013).
38. Cécile Sarabian and Andrew MacIntosh (2015); Sarabian et al. (2018), and Valerie Curtis (2014).
39. Jane Goodall (1986a), p. 466.
40. Jane Goodall (1986b).
41. James Anderson et al. (2017).
42. Andrew Ortony and Terence Turner (1990); Liah Greenfeld (2013).
43. Piera Filippi et al. (2017).
44. Robert Sapolsky (2017), p. 569.

CHAPTER 5: WILL TO POWER

1. *Mother Jones,* March 3, 2016.
2. Mauk Mulder (1977).
3. Ina Fried (September 5, 2005), *CNET News.*
4. Muzafer Sherif et al. (1954).
5. Jane Goodall (1990).
6. Toshisada Nishida (1996).
7. Jill Pruetz (2017).
8. Laurence Gesquiere et al. (2011).
9. Winston Churchill (1924).
10. Coren Apicella et al. (2012).
11. Richard Wrangham and Dale Peterson (1996), p. 63.
12. Michael Wilson et al. (2014).
13. Barbara Fruth and Gottfried Hohmann (2018).
14. Jingzhi Tan, Dan Ariely, and Brian Hare (2017).
15. James Rilling (2011).
16. Douglas Fry (2013).
17. John Horgan (2014).
18. Takeshi Furuichi (1997, 2011).

19. Takayoshi Kano (1992).
20. Nahoko Tokuyama and Takeshi Furuichi (2017).
21. Takeshi Furuichi (1997).
22. www.cnn.com/2016/01/12/europe/putin-merkel-scared-dog.

CHAPTER 6: EMOTIONAL INTELLIGENCE

1. Antonio Damasio (1994), p. 49-50.
2. www.sacred-texts.com/chr/thomas.htm.
3. Mark O'Connell (2015), Alan Jasanoff (2018).
4. Daniel Goleman (1995), Peter Salovey et al. (2003).
5. Pyotr Kropotkin (1902), p. 6.
6. Lee Dugatkin (2011).
7. Richard Easterlin (1974).
8. We captured the experiment on video a decade later, when it became an instant Internet hit with over a hundred million viewers. Seeing our test chamber, some commentators decried the conditions, thinking that the monkeys lived like this all the time. However, they'd be in the chamber for only half an hour before returning to their group: www.youtube.com/watch?v=meiU6TxysCg.
9. Sarah Brosnan and Frans de Waal (2003b, 2014).
10. evonomics.com/scientists-discover-what-economists-never-found-humans.
11. Friederike Range et al. (2008), Jennifer Essler et al. (2017).
12. Irene Pepperberg (2008), p. 153.
13. Sue Savage-Rumbaugh, interview by de Waal (1997), p. 41.
14. Rachel Sherman (2017).
15. Michael Alvard (2004).
16. Darby Proctor et al. (2013).
17. John Rawls (1972), p. 530.
18. Helmut Schoeck (1987), see also George Walsh (1992).
19. Christophe Boesch (1994).
20. Richard Wilkinson (2001).
21. Ernst Haeckel (1884), p. 238.
22. Harry Frankfurt (1971).
23. Harry Frankfurt (1971), p. 17.
24. Harry Frankfurt (2005), p. 63.
25. Carel van Schaik et al. (2013).

26. Michael Beran (2002).
27. Adrienne Koepke, Suzanne Gray, and Irene Pepperberg (2015).
28. Ted Evans and Michael Beran (2007).
29. Cécile Fruteau, Eric van Damme, and Ronald Noë (2013).
30. Winthrop and Luella Kellogg (1933).
31. Roy Baumeister (2008), p. 16.
32. Tomoko Sakai et al. (2012).
33. Suzana Herculano-Houzel (2009); Robert Barton and Chris Venditti (2013).
34. Jim Coan, Hillary Schaefer, and Richard Davidson (2006); Pavel Goldstein et al. (2018).
35. Marlise Hofer et al. (2018).
36. Zanna Clay and Frans de Waal (2013).
37. Nim Tottenham et al. (2010).
38. www.bbc.com/news/magazine-22987447.
39. Kathrin Wagner et al. (2015).

CHAPTER 7: SENTIENCE

1. Franz Kafka (1917). *A Report to an Academy.* www.kafka.org.
2. Joshua Plotnik, Diana Reiss, and Frans de Waal (2006).
3. Suzana Herculano-Houzel et al. (2014).
4. Martha Nussbaum (2001); Mark Rowlands (2009); Peter Godfrey-Smith (2016); Kristin Andrews and Jacob Beck (2018).
5. Craig Stanford (2001).
6. Loren Cordain et al. (2000).
7. Richard Wrangham (2009); Suzana Herculano-Houzel (2016).
8. https://faunalytics.org/wp-content/uploads/2015/06/Faunalytics_Current-Former-Vegetarians_Full-Report.pdf.
9. James Shapiro (2011), p. 143.
10. Antonio Damasio (1999), p. 138.
11. Heidi Appel and Rex Cocroft (2014).
12. Ken Yokawa et al. (2017).
13. Barry Magee and Robert Elwood (2013).
14. On chimera sentience: www.inverse.com/article/26995.
15. Kristin Hillman and David Bilkey (2010); Thomas Hills and Stephen Butterfill (2015).
16. David Premack and Ann Premack (1994).

17. Josep Call (2004).
18. Stanislas Dehaene and Lionel Naccache (2001).
19. John Garcia et al. (1955).
20. Danielle Panoz-Brown et al. (2018).
21. Karline Janmaat et al. (2014).
22. Nicola Clayton and Anthony Dickinson (1998).
23. David Smith et al. (1995).
24. Robert Hampton (2001).
25. Lisa Feldman Barrett (2016).
26. Lauri Nummenmaa et al. (2014).
27. Jaak Panksepp (1998).
28. John B. Watson (1913).
29. William James (1890), p. 148.
30. Promotional text on the jacket of Kenneth Miller (2018).
31. René Descartes (1633), p. 108.
32. Peter van Leeuwen et al. (2009).
33. Jaak Panksepp (2005), p. 31.
34. Ajay Mathuru et al. (2012).
35. Julian Pittman and Angelo Piato (2017).
36. Victoria Braithwaite (2010); Lynne Sneddon (2003); Sneddon, Braithwaite, and Michael Gentle (2003).
37. David Chamberlain (1991).
38. Menno Schilthuizen (2018).

CHAPTER 8: CONCLUSION

1. Sandra Gadanho and John Hallam (2001), p. 50.
2. Michael Arbib and Jean Marc Fellous (2004).

BIBLIOGRAPHY

Adang, O. 1999. *De Machtigste Chimpansee van Nederland*. Amsterdam: Nieuwezijds.

Alexander, R. D. 1986. Ostracism and indirect reciprocity: The reproductive significance of humor. *Ethology and Sociobiology* 7:253–70.

Alvard, M. 2004. The Ultimatum Game, fairness, and cooperation among big game hunters. In *Foundations of Human Sociality: Ethnography and Experiments from Fifteen Small-Scale Societies*, ed. J. Henrich et al., 413–35. London: Oxford University Press.

Anderson, J. R., et al. 2017. Third-party social evaluations of humans by monkeys and dogs. *Neuroscience and Biobehavioral Reviews* 82:95–109.

Anderson, J. R., A. Gillies, and L. C. Lock. 2010. Pan thanatology. *Current Biology* 20:R349–R351.

Andrew, R. J. 1963. The origin and evolution of the calls and facial expressions of the primates. *Behaviour* 20:1–109.

Andrews, K., and Beck, J. 2018. *The Routledge Handbook of Philosophy of Animal Minds*. Oxford: Routledge.

Apicella, C. L., F. W. Marlowe, J. H. Fowler, and N. A. Christakis. 2012. Social networks and cooperation in hunter-gatherers. *Nature* 481:497–501.

Appel, H. M., and R. B. Cocroft. 2014. Plants respond to leaf vibrations caused by insect herbivore chewing. *Oecologia* 175:1257–66.

Arbib, M. A., and J. M. Fellous. 2004. Emotions: From brain to robot. *Trends in Cognitive Sciences* 8:554–61.

Aureli, F., R. Cozzolino, C. Cordischi, and S. Scucchi. 1992. Kin-oriented redirection among Japanese macaques: An expression of a revenge system? *Animal Behaviour* 44:283–91.

Bailey, M. B. 1986. Every animal is the smartest: Intelligence and the ecological niche. In *Animal Intelligence,* ed. R. Hoage and L. Goldman, 105–13. Washington, DC: Smithsonian Institution Press.

Baron-Cohen, S. 2005. Autism—'autos': Literally, a total focus on the self? In *The Lost Self: Pathologies of the Brain and Identity,* ed. T. E. Feinberg and J. P. Keenan, 166–80. Oxford: Oxford University Press.

Barrett, L. F. 2016. Are emotions natural kinds? *Perspectives on Psychological Science* 1:28–58.

Bartal, I. B.-A., et al. 2016. Anxiolytic treatment impairs helping behavior in rats. *Frontiers in Psychology* 7:850.

Bartal, I. B.-A., J. Decety, and P. Mason. 2011. Empathy and pro-social behavior in rats. *Science* 334:1427–30.

Barton, R. A., and C. Venditti. 2013. Human frontal lobes are not relatively large. *Proceedings of the National Academy of Sciences USA* 110:9001–9006.

Baumeister, R. F. 2008. Free will in scientific psychology. *Perspectives on Psychological Science* 3:14–19.

Bekoff, M. 1972. The development of social interaction, play, and metacommunication in mammals: An ethological perspective. *Quarterly Review of Biology* 47:412–34.

Beran, M. J. 2002. Maintenance of self-imposed delay of gratification by four chimpanzees (*Pan troglodytes*) and an orangutan (*Pongo pygmaeus*). *Journal of General Psychology* 129:49–66.

Berns, G. S., A. Brooks, and M. Spivak. 2013. Replicability and heterogeneity of awake unrestrained canine fMRI responses. *PLoS ONE* 8:e81698.

Biro, D., T. Humle, K. Koops, C. Sousa, M. Hayashi, and T. Matsuzawa. 2010. Chimpanzee mothers at Bossou, Guinea, carry the mummified remains of their dead infants. *Current Biology* 20:R351–R352.

Bloom, P. 2016. *Against Empathy: The Case for Rational Compassion.* New York: Ecco.

Boesch, C. 1994. Cooperative hunting in wild chimpanzees. *Animal Behaviour* 48:653–67.

Bosch, O. J., et al. 2009. The CRF System mediates increased passive stress-coping behavior following the loss of a bonded partner in a monogamous rodent. *Neuropsychopharmacology* 34:1406–15.

Braithwaite, V. 2010. *Do Fish Feel Pain?* Oxford: Oxford University Press.

Brosnan, S. F., and F. B. M. de Waal. 2003a. Regulation of vocal output by chimpanzees finding food in the presence or absence of an audience. *Evolution of Communication* 4:211–24.

————. 2003b. Monkeys reject unequal pay. *Nature* 425:297–99.

————. 2014. The evolution of responses to (un)fairness. *Science* 346:314–322.

Brotcorne, F., et al. 2017. Intergroup variation in robbing and bartering by long-tailed macaques at Uluwatu Temple (Bali, Indonesia). *Primates* 58:505–16.

Buchanan, T. W., S. L. Bagley, R. B. Stansfield, and S. D. Preston. 2012. The empathic, physiological resonance of stress. *Social Neuroscience* 7:191–201.

Burkett, J., et al. 2016. Oxytocin-dependent consolation behavior in rodents. *Science* 351:375–78.

Burrows, A. M., B. M. Waller, L. A. Parr, and C. J. Bonar. 2006. Muscles of facial expression in the chimpanzee (*Pan troglodytes*): Descriptive, comparative and phylogenetic contexts. *Journal of Anatomy* 208:153–67.

Calcutt, S. E., T. L. Rubin, J. Pokorny, and F. B. M. de Waal. 2017. Discrimination of emotional facial expressions by tufted capuchin monkeys (*Sapajus apella*). *Journal of Comparative Psychology* 131:40–49.

Call, J. 2004. Inferences about the location of food in the great apes. *Journal of Comparative Psychology* 118:232–41.

Campbell, M. W., and F. B. M. de Waal. 2011. Ingroup-outgroup bias in contagious yawning by chimpanzees supports link to empathy. *PloS ONE* 6:e18283.

Caruana, F., et al. 2011. Emotional and social behaviors elicited by electrical stimulation of the insula in the macaque monkey. *Current Biology* 21:195–99.

Chamberlain, D. B. 1991. Babies don't feel pain: A century of denial in medicine. Lecture at the 2nd International Symposium on Circumcision, San Francisco, CA.

Chen, P. Z., R. L. Carrasco, and P. K. L. Ng. 2017. Mangrove crab uses victory display to "browbeat" losers from re-initiating a new fight. *Ethology* 123:981–88.

Chester, D. S., and C. N. DeWall. 2017. Combating the sting of rejection with the pleasure of revenge: A new look at how emotion shapes aggression. *Journal of Personality and Social Psychology* 112:413–30.

Churchill, W. S. 1924. Shall we commit suicide? *Nash's Pall Mall Magazine*.

Churchland, P. S. 2011. *Braintrust: What Neuroscience Tells Us about Morality.* Princeton, NJ: Princeton University Press.

Clay, Z., and F. B. M. de Waal. 2013. Development of socio-emotional competence in bonobos. *Proceedings of the National Academy of Sciences USA* 110:18121–26.

Clayton, N. S., and A. Dickinson. 1998. Episodic-like memory during cache recovery by scrub jays. *Nature* 395:272–74.

Coan, J. A., H. S. Schaefer, and R. J. Davidson. 2006. Lending a hand: Social regulation of the neural response to threat. *Psychological Science* 17:1032–39.

Coe, C. L., and L. A. Rosenblum. 1984. Male dominance in the bonnet macaque: A malleable relationship. In *Social Cohesion: Essays Toward a Sociophysiological Perspective,* ed. P. R. Barchas and S. P. Mendoza, 31–63. Westport, CT: Greenwood.

Cordain, L., et al. 2000. Plant-animal subsistence ratios and macronutrient energy estimations in worldwide hunter-gatherer diets. *American Journal of Clinical Nutrition* 71:682–92.

Crick, F. 1995. *The Astonishing Hypothesis: The Scientific Search for the Soul.* New York: Scribner.

Curtis, V. A. 2014. Infection-avoidance behaviour in humans and other animals. *Trends in Immunology* 35:457–64.

Custance, D., and J. Mayer. 2012. Empathic-like responding by domestic dogs (*Canis familiaris*) to distress in humans: An exploratory study. *Animal Cognition* 15:851–59.

Damasio, A. R. 1994. *Descartes' Error: Emotion, Reason, and the Human Brain.* New York: Putnam.

———. 1999. *The Feeling of What Happens: Body and Emotion in the Making of Consciousness.* New York: Harcourt.

Darwin, C. 1987. *The Correspondence of Charles Darwin,* vol. 2: *1837–1843.* Ed. F. Burkhardt and S. Smith. Cambridge: Cambridge University Press.

———. 1998 [orig. 1872]. *The Expression of the Emotions in Man and Animals.* New York: Oxford University Press.

Davila Ross, M., S. Menzler, and E. Zimmermann. 2007. Rapid facial mimicry in orangutan play. *Biology Letters* 4:27–30.

de Montaigne, M. 2003 [orig. 1580]. *The Complete Essays.* London: Penguin.

de Waal, F. B. M. 1982. *Chimpanzee Politics.* London: Jonathan Cape.

———. 1986. The brutal elimination of a rival among captive male chimpanzees. *Ethology and Sociobiology* 7:237–51.

———. 1989. *Peacemaking Among Primates.* Cambridge, MA: Harvard University Press.

———. 1997a. The chimpanzee's service economy: Food for grooming. *Evolution and Human Behavior* 18:375–86.

———. 1997b. *Bonobo: The Forgotten Ape.* Berkeley: University of California Press.

———. 2007 [orig. 1982]. *Chimpanzee Politics: Power and Sex among Apes.* Baltimore: Johns Hopkins University Press.

———. 2008. Putting the altruism back into altruism: The evolution of empathy. *Annual Review of Psychology* 59:279–300.

———. 2011. What is an animal emotion? *The Year in Cognitive Neuroscience, Annals of the New York Academy of Sciences* 1224:191–206.

———. 2013. *The Bonobo and the Atheist: In Search of Humanism Among the Primates.* New York: Norton.

———. 2016. *Are We Smart Enough to Know How Smart Animals Are?* New York: Norton.

de Waal, F. B. M., and L. M. Luttrell. 1985. The formal hierarchy of rhesus monkeys: An investigation of the bared-teeth display. *American Journal of Primatology* 9:73–85.

———. 1988. Mechanisms of social reciprocity in three primate species: Symmetrical relationship characteristics or cognition? *Ethology and Sociobiology* 9:101–18.

de Waal, F. B. M., and J. Pokorny. 2008. Faces and behinds: Chimpanzee sex perception. *Advanced Science Letters* 1:99–103.

Dehaene, S., and L. Naccache. 2001. Towards a cognitive neuroscience of consciousness: Basic evidence and a workspace framework. *Cognition* 79:1–37.

Descartes, R. 2003 [orig. 1633]. *Treatise of Man.* Paris: Prometheus.

Dimberg, U., P. Andréasson, and M. Thunberg. 2011. Emotional empathy and facial reactions to facial expressions. *Journal of Psychophysiology* 25:26–31.

Dimberg, U., M. Thunberg, and K. Elmehed. 2000. Unconscious facial reactions to emotional facial expressions. *Psychological Science* 11:86–89.

Douglas, C., et al. 2012. Environmental enrichment induces optimistic cognitive biases in pigs. *Applied Animal Behaviour Science* 139:65–73.

Dugatkin, L. A. 2011. *The Prince of Evolution: Peter Kropotkin's Adventures in Science and Politics.* CreateSpace.

Easterlin, R. 1974. Does economic growth improve the human lot? In *Nations and Households in Economic Growth: Essays in Honor of Moses Abramovitz.* Ed. M. Abramovitz, P. David, and M. Reder, 89–125. New York: Academic Press.

Eibl-Eibesfeldt, I. 1973. *Der vorprogrammierte Mensch: Das Ererbte als bestimmender Faktor im menschlichen Verhalten.* Vienna: Verlag Fritz Molden.

Ekman, P. 1998. Afterword: Universality of emotional expression? A personal history of the dispute. In *Darwin,* ed. P. Ekman, 363–93. New York: Oxford University Press.

Ekman, P., and W. V. Friesen. 1971. Constants across cultures in the face and emotion. *Journal of Personality and Social Psychology* 17:124–29.

Essler, J. L., W. V. Marshall-Pescini, and F. Range, 2017. Domestication does not explain the presence of inequity aversion in dogs. *Current Biology* 27:1861–65.

Evans, T. A., and M. J. Beran. 2007. Chimpanzees use self-distraction to cope with impulsivity. *Biology Letters* 3:599–602.

Fehr, E., H. Bernhard, and B. Rockenbach, 2008. Egalitarianism in young children. *Nature* 454:1079–83.

Fessler, D. M. T. 2004. Shame in two cultures: Implications for evolutionary approaches. *Journal of Cognition and Culture* 4:207–62.

Filippi, P. et al. 2017. Humans recognize emotional arousal in vocalizations across all classes of terrestrial vertebrates: Evidence for acoustic universals. *Proceedings of the Royal Society* B 284:20170990.

Finlayson, K., J. F. Lampe, S. Hintze, H. Würbel, and L. Melotti. 2016. Facial indicators of positive emotions in rats. *PLoS ONE* 11:e0166446.

Flack, J. C., L. A. Jeannotte, and F. B. M. de Waal. 2004. Play signaling and the perception of social rules by juvenile chimpanzees. *Journal of Comparative Psychology* 118:149–59.

Foerster, S., et al. 2016. Chimpanzee females queue but males compete for social status. *Scientific Reports* 6:35404.

Fouts, R., and T. Mills. 1997. *Next of Kin*. New York: Morrow.

Frankfurt, H. G. 1971. Freedom of the will and the concept of a person. *Journal of Philosophy* 68:5–20.

———. 2005. *On Bullshit*. Princeton, NJ: Princeton University Press.

Fruteau, C., E. van Damme, and R. Noë. 2013. Vervet monkeys solve a multiplayer "forbidden circle game" by queuing to learn restraint. *Current Biology* 23:665–70.

Fruth, B., and G. Hohmann. 2018. Food sharing across borders: First observation of intercommunity meat sharing by bonobos at LuiKotale, DRC. *Human Nature* 29:91–103.

Fry, D. P. 2013. *War, Peace, and Human Nature: The Convergence of Evolutionary and Cultural Views*. Oxford: Oxford University Press.

Furuichi, T. 1997. Agonistic interactions and matrifocal dominance rank of wild bonobos (*Pan paniscus*) at Wamba. *International Journal of Primatology* 18:855–75.

———. 2011. Female contributions to the peaceful nature of bonobo society. *Evolutionary Anthropology* 20:131–42.

Gadanho, S. C., and J. Hallam. 2001. Robot learning driven by emotions. *Adaptive Behavior* 9:42–64.

Garcia, J., D. J. Kimeldorf, and R. A. Koelling. 1955. Conditioned aversion to saccharin resulting from exposure to gamma radiation. *Science* 122:157–58.

Gazzaniga, M. S. 2008. *Human: The Science Behind What Makes Your Brain Unique.* New York: Ecco.

Gesquiere, L. R., et al. 2011. Life at the top: Rank and stress in wild male baboons. *Science* 333:357–60.

Ghiselin, M. 1974. *The Economy of Nature and the Evolution of Sex.* Berkeley: University of California Press.

Godfrey-Smith, P. 2016. *Other Minds: The Octopus, the Sea, and the Deep Origins of Consciousness.* New York: Farrar, Strauss and Giroux.

Goldstein, P., I. Weissman-Fogel, G. Dumas, and S. G. Shamay-Tsoory. 2018. Brain-to-brain coupling during handholding is associated with pain reduction. *Proceedings of the National Academy of Sciences USA* 115:201703643.

Goleman, D. 1995. *Emotional Intelligence.* New York: Bantam.

Goodall, J. 1986a. *The Chimpanzees of Gombe: Patterns of Behavior.* Cambridge, MA: Belknap.

———. 1986b. Social rejection, exclusion, and shunning among the Gombe chimpanzees. *Ethology and Sociobiology* 7:227–36.

———. 1990. *Through a Window: My Thirty Years with the Chimpanzees of Gombe.* Boston: Houghton Mifflin.

Grandin, T., and C. Johnson. 2009. *Animals Make Us Human: Creating the Best Life for Animals.* Boston: Houghton Mifflin.

Greenfeld, L. 2013. Are human emotions universal? *Psychology Today.*

Haeckel, E. 2012 [orig. 1884]. *The History of Creation, Or the Development of the Earth and its Inhabitants by the Action of Natural Causes,* vol. 1. Project Gutenberg.

Hampton, R. R. 2001. Rhesus monkeys know when they remember. *Proceedings of the National Academy of Sciences USA* 98:5359–62.

Hare, B., and S. Kwetuenda. 2010. Bonobos voluntarily share their own food with others. *Current Biology* 20:R230–R231.

Hebb, D. O. 1946. Emotion in man and animal: An analysis of the intuitive processes of recognition. *Psychological Review* 53:88–106.

Herculano-Houzel, S. 2009. The human brain in numbers: A linearly scaled-up primate brain. *Frontiers in Human Neuroscience* 3:1–11.

———. 2016. *The Human Advantage: A New Understanding of How Our Brain Became Remarkable.* Cambridge, MA: MIT Press.

Herculano-Houzel, S., et al. 2014. The elephant brain in numbers. *Frontiers in Neuroanatomy* 8:46.

Hillman, K. L., and D. K. Bilkey. 2010. Neurons in the rat Anterior Cingulate Cortex dynamically encode cost–benefit in a spatial decision-making task. *Journal of Neuroscience* 30:7705–13.

Hills, T. T., and S. Butterfill. 2015. From foraging to autonoetic consciousness: The primal self as a consequence of embodied prospective foraging. *Current Zoology* 61:368–81.

Hobaiter, C., and R. W. Byrne. 2010. Able-bodied wild chimpanzees imitate a motor procedure used by a disabled individual to overcome handicap. *PLoS ONE* 5:e11959.

Hockings, K. J., et al. 2007. Chimpanzees share forbidden fruit. *PLoS ONE* 2:e886.

Hofer, M. K., H. K. Collins, A. V. Whillans, and F. S. Chen. 2018. Olfactory cues from romantic partners and strangers influence women's responses to stress. *Journal of Personality and Social Psychology* 114:1–9.

Hoffman, M. L. 1981. Is altruism part of human nature? *Journal of Personality and Social Psychology* 40:121–37.

Horgan, J. 2014. Thanksgiving and the slanderous myth of the savage savage. *Scientific American Cross-Check Blog.*

Horner, V., and F. B. M. de Waal. 2009. Controlled studies of chimpanzee cultural transmission. *Progress in Brain Research* 178:3–15.

Horner, V., D. J. Carter, M. Suchak, and F. B. M. de Waal, 2011. Spontaneous prosocial choice by chimpanzees. *Proceedings of the Academy of Sciences USA* 108:13847–51.

Horowitz, A. 2009. *Inside of a Dog: What Dogs See, Smell, and Know.* New York: Scribner.

Hrdy, S. B. 2009. *Mothers and Others: The Evolutionary Origins of Mutual Understanding.* Cambridge, MA: Belknap.

James, W. 1950 [orig. 1890]. *The Principles of Psychology.* New York: Dover.

Janmaat, K. R. L., L. Polansky, S. D. Ban, and C. Boesch. 2014. Wild chimpanzees plan their breakfast time, type, and location. *Proceedings of the National Academy of Sciences USA* 111:16343–48.

Jasanoff, A. 2018. *The Biological Mind: How Brain, Body, and Environment Collaborate to Make Us Who We Are.* New York: Basic Books.

Kaburu, S. S. K., S. Inoue, and N. E. Newton-Fisher. 2013. Death of the alpha: Within-community lethal violence among chimpanzees of the Mahale Mountains National Park. *American Journal of Primatology* 75:789–97.

Kaminski, J., et al. 2017. Human attention affects facial expressions in domestic dogs. *Scientific Reports* 7:12914.

Kano, T. 1992. *The Last Ape: Pygmy Chimpanzee Behavior and Ecology*. Stanford, CA: Stanford University Press.

Kellogg, W. N., and L. A. Kellogg. 1967 [orig. 1933]. *The Ape and the Child: A Study of Environmental Influence upon Early Behavior*. New York: Hafner.

King, B. J. 2013. *How Animals Grieve*. Chicago: University of Chicago Press.

Koepke, A. E., S. L. Gray, and I. M. Pepperberg. 2015. Delayed gratification: A grey parrot (*Psittacus erithacus*) will wait for a better reward. *Journal of Comparative Psychology* 129:339–46.

Kraus, M. W., and T. W. Chen. 2013. A winning smile? Smile intensity, physical dominance, and fighter performance. *Emotion* 13:270–79.

Kropotkin, P. 2009 [orig. 1902]. *Mutual Aid: A Factor of Evolution*. New York: Cosimo.

Ladygina-Kohts, N. N. 2002 [orig. 1935]. *Infant Chimpanzee and Human Child: A Classic 1935 Comparative Study of Ape Emotions and Intelligence*. Ed. F. B. M. de Waal. Oxford: Oxford University Press.

Lahr, J. 2000. *Dame Edna Everage and the Rise of Western Civilisation: Backstage with Barry Humphries,* 2nd ed. Berkeley: University of California Press.

Langford, D. J., et al. 2010. Coding of facial expressions of pain in the laboratory mouse. *Nature Methods* 7:447–49.

Lazarus, R., and B. Lazarus. 1994. *Passion and Reason*. New York: Oxford University Press.

LeDoux, J. E. 2014. Coming to terms with fear. *Proceedings of the National Academy of Sciences USA* 111:2871–78.

Leuba, J. H. 1928. Morality among the animals. *Harper's Monthly* 937:97–103.

Limbrecht-Ecklundt, K., et al. 2013. The effect of forced choice on facial emotion recognition: A comparison to open verbal classification of emotion labels. *GMS Psychosocial Medicine* 10.

Lindegaard, M. R., et al. 2017. Consolation in the aftermath of robberies resembles post-aggression consolation in chimpanzees. *PLoS ONE* 12:e0177725.

Lipps, T. 1903. Einfühlung, innere Nachahmung und Organenempfindungen. *Archiv für die gesamte Psychologie* 1:465–519.

Lorenz, K. 1960. *So kam der Mensch auf den Hund*. Vienna: Borotha-Schoeler.

———. 1966. *On Aggression*. New York: Harcourt.

———. 1980. Tiere sind Gefühlsmenschen. *Der Spiegel* 47:251–64.

Magee, B., and R. E. Elwood. 2013. Shock avoidance by discrimination learning in the shore crab (*Carcinus maenas*) is consistent with a key criterion for pain. *Journal of Experimental Biology* 216:353–58.

Maslow, A. H. 1936. The role of dominance in the social and sexual behavior of infra-human primates: I. Observations at Vilas Park Zoo. *Journal of Genetic Psychology* 48:261–77.

Masson, J. M., and S. McCarthy. 1995. *When Elephants Weep: The Emotional Lives of Animals.* New York: Delacorte.

Mathuru, A. S., et al. 2012. Chondroitin fragments are odorants that trigger fear behavior in fish. *Current Biology* 22:538–44.

Matsuzawa, T. 2011. What is uniquely human? A view from comparative cognitive development in humans and chimpanzees. In *The Primate Mind,* ed. F. B. M. de Waal and P. F. Ferrari, 288–305. Cambridge, MA: Harvard University Press.

McConnell, P. 2005. *For the Love of a Dog.* New York: Ballantine Books.

McFarland, D. 1987. *The Oxford Companion to Animal Behaviour.* Oxford: Oxford University Press.

Mendl, M., O. H. P. Burman, and E. S. Paul. 2010. An integrative and functional framework for the study of animal emotion and mood. *Proceeding of the Royal Society B* 277:2895–904.

Michl, P., et al. 2014. Neurobiological underpinnings of shame and guilt: A pilot fMRI study. *Social Cognitive and Affective Neuroscience* 9:150–57.

Miller, K. R. 2018. *The Human Instinct: How We Evolved to Have Reason, Consciousness, and Free Will.* New York: Simon and Schuster.

Mogil, J. S. 2015. Social modulation of and by pain in humans and rodents. *PAIN* 156:S35–S41.

Mulder, M. 1977. *The Daily Power Game.* Amsterdam: Nijhoff.

Nagasaka, Y. et al. 2013. Spontaneous synchronization of arm motion between Japanese macaques. *Scientific Reports* 3:1151.

Neal, D. T., and T. L. Chartrand. 2011. Amplifying and dampening facial feedback modulates emotion perception accuracy. *Social Psychological and Personality Science* 2:673–78.

Nishida, T. 1996. The death of Ntologi: The unparalleled leader of M Group. *Pan Africa News* 3:4.

Norscia, I., and E. Palagi. 2011. Yawn contagion and empathy in *Homo sapiens. PloS ONE* 6:e28472.

Nowak, M., and R. Highfield. 2011. *SuperCooperators: Altruism, Evolution, and Why We Need Each Other to Succeed.* New York: Free Press.

Nummenmaa, L., E. Glerean, R. Hari, and J. K. Hietanen. 2014. Bodily maps of emotions. *Proceedings of the National Academy of Sciences USA* 111:646–51.

Nussbaum, M. 2001. *Upheavals of Thought: The Intelligence of Emotions.* Cambridge: Cambridge University Press.

O'Brien, E., S. H. Konrath, D. Grühn, and A. L. Hagen. 2013. Empathic concern and perspective taking: Linear and quadratic effects of age across the adult life span. *Journals of Gerontology, Series B: Psychological Sciences and Social Sciences* 68:168–75.

O'Connell, C. 2015. *Elephant Don: The Politics of a Pachyderm Posse.* Chicago: University of Chicago Press.

O'Connell, M. 2017. *To Be a Machine.* London: Granta.

Ortony, A., and T. J. Turner. 1990. What's basic about basic emotions? *Psychological Review* 97:315–31.

Osvath, M., and H. Osvath. 2008. Chimpanzee (*Pan troglodytes*) and orangutan (*Pongo abelii*) forethought: Self-control and pre-experience in the face of future tool use. *Animal Cognition* 11:661–74.

Panksepp, J. 1998. *Affective Neuroscience: The Foundations of Human and Animal Emotions.* New York: Oxford University Press.

———. 2005. Affective consciousness: Core emotional feelings in animals and humans. *Consciousness and Cognition* 14:30–80.

Panksepp, J., and Burgdorf, J. 2003. "Laughing" rats and the evolutionary antecedents of human joy? *Physiology and Behavior* 79:533–47.

Panoz-Brown, D., et al. 2018. Replay of episodic memories in the rat. *Current Biology* 28:1–7.

Parr, L. A. 2001. Cognitive and physiological markers of emotional awareness in chimpanzees (*Pan troglodytes*). *Animal Cognition* 4:223–29.

Parr, L. A., M. Cohen, and F. B. M. de Waal. 2005. Influence of social context on the use of blended and graded facial displays in chimpanzees. *International Journal of Primatology* 26:73–103.

Paukner, A., S. J. Suomi, E. Visalberghi, and P. F. Ferrari. 2009. Capuchin monkeys display affiliation toward humans who imitate them. *Science* 325:880–83.

Payne, K. 1998. *Silent Thunder: In the Presence of Elephants.* New York: Simon and Schuster.

Pepperberg, I. M. 2008. *Alex and Me.* New York: Collins.

Perry, S., et al. 2003. Social conventions in wild white-faced capuchin monkeys: Evidence for traditions in a neotropical primate. *Current Anthropology* 44:241–68.

Pinker, S. 2011. *The Better Angels of Our Nature: Why Violence Has Declined.* New York: Viking.

Pittman, J., and A. Piato. 2017. Developing zebrafish depression-related models. In *The Rights and Wrongs of Zebrafish: Behavioral Phenotyping of Zebrafish,* ed. A. V. Kalueff, 33–43. Cham: Springer.

Plotnik, J. M., and F. B. M. de Waal. 2014. Asian elephants (*Elephas maximus*) reassure others in distress. *PeerJ* 2:e278.

Plotnik, J. M., F. B. M. de Waal, and D. Reiss. 2006. Self-recognition in an Asian elephant. *Proceedings of the National Academy of Sciences USA* 103:17053–57.

Premack, D., and A. J. Premack. 1994. Levels of causal understanding in chimpanzees and children. *Cognition* 50:347–62.

Proctor, D., R. A. Williamson, F. B. M. de Waal, and S. F. Brosnan. 2013. Chimpanzees play the Ultimatum Game. *Proceedings of the National Academy of Sciences USA* 110:2070–75.

Proust, M. 1982. *Remembrance of Things Past,* 3 vols. New York: Vintage Press.

Provine, R. R. 2000. *Laughter: A Scientific Investigation.* New York: Viking.

Pruetz, J. D., et al. 2017. Intragroup lethal aggression in West African chimpanzees (*Pan troglodytes verus*): Inferred killing of a former alpha male at Fongoli, Senegal. *International Journal of Primatology* 38:31–57.

Range, F., L. Horn, Z. Viranyi, and L. Huber. 2008. The absence of reward induces inequity aversion in dogs. *Proceedings of the National Academy of Sciences USA* 106:340–45.

Rawls, J. 1972. *A Theory of Justice.* Oxford: Oxford University Press.

Rilling, J. K., et al. 2011. Differences between chimpanzees and bonobos in neural systems supporting social cognition. *Social Cognitive and Affective Neuroscience* 7:369–79.

Romero, T., M. A. Castellanos, and F. B. M. de Waal. 2010. Consolation as possible expression of sympathetic concern among chimpanzees. *Proceedings of the National Academy of Sciences* 107:12110–15.

Rowlands, M. 2009. *The Philosopher and the Wolf: Lessons from the Wild on Love, Death and Happiness.* New York: Pegasus.

Rozin, P., J. Haidt, and C. McCauley. 2000. Disgust. In *Handbook of Emotions,* ed. M. Lewis and S. M. Haviland-Jones, 637–53. New York: Guilford.

Sakai, T. et al. 2012. Fetal brain development in chimpanzees versus humans. *Current Biology* 22:R791–R792.

Salovey, P., M. Kokkonen, P. N. Lopes, and J. D. Mayer. 2003. Emotional intelligence. In *Feelings and Emotions: The Amsterdam Symposium,* eds. T. Manstead, N. Frijda, and A. Fischer, 321–340. Cambridge: Cambridge University Press.

Sanfey, A. G., J. K. Rilling, J. A. Aronson, L. E. Nystrom, and J. D. Cohen. 2003. The neural basis of economic decision-making in the ultimatum game. *Science* 300:1755–58.

Sapolsky, R. M. 2017. *Behave: The Biology of Humans at Our Best and Worst.* New York: Penguin.

Sarabian, C., and A. J. J. MacIntosh. 2015. Hygienic tendencies correlate with low geohelminth infection in free-ranging macaques. *Biology Letters* 11:20150757.

Sarabian, C., R. Belais, and A. J. J. MacIntosh. 2018. Feeding decisions under contamination risk in bonobos. *Philosophical Transactions of the Royal Society B* 373.

Sato, N., L. Tan, K. Tate, and M. Okada. 2015. Rats demonstrate helping behavior toward a soaked conspecific. *Animal Cognition* 18:1039–47.

Sauter, D. A., O. LeGuen, and D. B. M. Haun. 2011. Categorical perception of emotional facial expressions does not require lexical categories. *Emotion* 11:1479–83.

Scheele, D., et al. 2012. Oxytocin modulates social distance between males and females. *Journal of Neuroscience* 32:16074–79.

Schilder, M. B. H., et al. 1984. A quantitative analysis of facial expression in the plains zebra. *Zeitschrift für Tierpsychologie* 66:11–32.

Schilthuizen, M. 2018. *Darwin Comes to Town: How the Urban Jungle Drives Evolution.* New York: Picador.

Schneiderman, I., et al. 2012. Oxytocin during the initial stages of romantic attachment: Relations to couples' interactive reciprocity. *Psychoneuroendocrinology* 37:1277–85.

Schoeck, H. 1987. *Envy: A Theory of Social Behaviour.* Indianapolis: Liberty Fund.

Schwing, R., X. J. Nelson, A. Wein, and S. Parsons. 2017. Positive emotional contagion in a New Zealand parrot. *Current Biology* 27:R213–R214.

Shapiro, J. A. 2011. *Evolution: A View from the 21st Century.* Upper Saddle River, NJ: FT Press Science.

Sherif, M., et al. 1954. *Experimental study of positive and negative intergroup attitudes between experimentally produced groups: Robbers' Cave Study.* Norman: University of Oklahoma Press.

Sherman, R. 2017. *Uneasy Street: The Anxieties of Affluence.* Princeton, NJ: Princeton University Press.

Singer, T., B. Seymour, J. P. O'Doherty, K. E. Stephan, R. J. Dolan, and C. D. Frith. 2006. Empathic neural responses are modulated by the perceived fairness of others. *Nature* 439:466–69.

Skinner, B. F. 1965 [1953]. *Science and Human Behavior.* New York: Free Press.

Sliwa, J., and W. A. Freiwald. 2017. A dedicated network for social interaction processing in the primate brain. *Science* 356:745–49.

Smith, A. 1937 [orig. 1759]. *A Theory of Moral Sentiments.* New York: Modern Library.

———. 1982 [orig. 1776]. *An Inquiry into the Nature and Causes of the Wealth of Nations.* Indianapolis: Liberty Classics.

Smith, J. D., J. Schull, J. Strote, K. McGee, R. Egnor, and L. Erb. 1995. The uncertain response in the bottlenosed dolphin (*Tursiops truncatus*). *Journal of Experimental Psychology: General* 124:391–408.

Sneddon, L. U. 2003. Evidence for pain in fish: The use of morphine as an analgesic. *Applied Animal Behaviour Science* 83:153–62.

Sneddon, L. U., V. A. Braithwaite, and M. J. Gentle. 2003. Do fishes have nociceptors? Evidence for the evolution of a vertebrate sensory system. *Proceedings of the Royal Society, London B* 270:1115–21.

Springsteen, B. 2016. *Born to Run.* New York: Simon and Schuster.

Stanford, C. B. 2001. *Significant Others: The Ape-Human Continuum and the Quest for Human Nature.* New York: Basic Books.

Stomp, M., et al. 2018. An unexpected acoustic indicator of positive emotions in horses. *PLoS ONE* 13:e0197898.

Suchak, M., and F. B. M. de Waal. 2012. Monkeys benefit from reciprocity without the cognitive burden. *Proceedings of the National Academy of Sciences USA* 109:15191–96.

Tan, J., and B. Hare. 2013. Bonobos share with strangers. *PloS ONE* 8:e51922.

Tan, J., D. Ariely, and B. Hare. 2017. Bonobos respond prosocially toward members of other groups. *Scientific Reports* 7:14733.

Tangney, J., and R. Dearing. 2002. *Shame and Guilt.* New York: Guilford.

Teleki, G. 1973. Group response to the accidental death of a chimpanzee in Gombe National Park, Tanzania. *Folia primatologica* 20:81–94.

Tinklepaugh, O. L. 1928. An experimental study of representative factors in monkeys. *Journal of Comparative Psychology* 8:197–236.

Tokuyama, N., and T. Furuichi. 2017. Do friends help each other? Patterns of female coalition formation in wild bonobos at Wamba. *Animal Behaviour* 119:27–35.

Tolstoy, L. 1975 [orig. 1904]. *The Lion and the Dog.* Moscow: Progress Publishers.

Tottenham, N., et al. 2010. Prolonged institutional rearing is associated with atypically large amygdala volume and difficulties in emotion regulation. *Developmental Science* 13:46–61.

Tracy, J. 2016. *Take Pride: Why the Deadliest Sin Holds the Secret to Human Success.* New York: Houghton.

Tracy, J. L., and D. Matsumoto. 2008. The spontaneous expression of pride and shame: Evidence for biologically innate nonverbal displays. *Proceedings of the National Academy of Sciences USA* 105:11655–60.

Troje, N. F. 2002. Decomposing biological motion: A framework for analysis and synthesis of human gait patterns. *Journal of Vision* 2:371–87.

Tybur, J. M., D. Lieberman, and V. Griskevicius. 2009. Microbes, mating, and morality: Individual differences in three functional domains of disgust. *Journal of Personality and Social Psychology* 97:103–22.

van de Waal, E., C. Borgeaud, and A. Whiten. 2013. Potent social learning and conformity shape a wild primate's foraging decisions. *Science* 340:483–85.

van Hooff, J. A. R. A. M. 1972. A comparative approach to the phylogeny of laughter and smiling. In *Non-verbal Communication,* ed. R. Hinde, 209–241. Cambridge: Cambridge University Press.

van Leeuwen, P., et al. 2009. Influence of paced maternal breathing on fetal-maternal heart rate coordination. *Proceedings of the National Academy of Sciences USA* 106:13661–66.

van Schaik, C. P., L. Damerius, and K. Isler. 2013. Wild orangutan males plan and communicate their travel direction one day in advance. *PLoS ONE* 8:e74896.

van Wyhe, J., P. C. Kjærgaard. 2015. Going the whole orang: Darwin, Wallace and the natural history of orangutans. *Studies in History and Philosophy of Biological and Biomedical Sciences* 51:53–63.

Vianna, D. M., and P. Carrive. 2005. Changes in cutaneous and body temperature during and after conditioned fear to context in the rat. *European Journal of Neuroscience* 21:2505–12.

Wagner, K., et al. 2015. Effects of mother versus artificial rearing during the first 12 weeks of life on challenge responses of dairy cows. *Applied Animal Behaviour Science* 164:1–11.

Walsh, G. V. 1992. Rawls and envy. *Reason Papers* 17:3–28.

Warneken, F., and M. Tomasello. 2014. Extrinsic rewards undermine altruistic tendencies in 20-month-olds. *Motivation Science* 1:43–48.

Wathan, J., et al. 2015. EquiFACS: The Equine Facial Action Coding System. *PLoS ONE* 10:e0131738.

Watson, J. B. 1913. Psychology as the behaviorist views it. *Psychological Review* 20:158–77.

Westermarck, E. 1912 [orig. 1908]. *The Origin and Development of the Moral Ideas.* Vol. 1. 2nd ed. London: Macmillan.

Wilkinson, R. 2001. *Mind the Gap.* New Haven, CT: Yale University Press.

Wilson, M. L., et al. 2014. Lethal aggression in Pan is better explained by adaptive strategies than human impacts. *Nature* 513:414–17.

Wispé, L. 1991. *The Psychology of Sympathy.* New York: Plenum.

Woodward, R., and C. Bernstein. 1976. *The Final Days.* New York: Simon and Schuster.

Wrangham, R. W. 2009. *Catching Fire: How Cooking Made Us Human.* New York: Basic Books.

Wrangham, R. W., and D. Peterson. 1996. *Demonic Males: Apes and the Evolution of Human Aggression.* Boston: Houghton Mifflin.

Yamamoto, S., T. Humle, and M. Tanaka. 2012. Chimpanzees' flexible targeted helping based on an understanding of conspecifics' goals. *Proceedings of the National Academy of Sciences USA* 109:3588–92.

Yerkes, R. M. 1941. Conjugal contrasts among chimpanzees. *Journal of Abnormal and Social Psychology* 36:175–99.

Yokawa, K., et al. 2017. Anaesthetics stop diverse plant organ movements, affect endocytic vesicle recycling and ROS homeostasis, and block action potentials in Venus flytraps. *Annals of Botany*: mcx155.

Young, L., and B. Alexander. 2012. *The Chemistry Between Us: Love, Sex, and the Science of Attraction.* New York: Current.

Zahn-Waxler, C., and M. Radke-Yarrow. 1990. The origins of empathic concern. *Motivation and Emotion* 14:107–30.

Zamma, K. 2002. A chimpanzee trifling with a squirrel: Pleasure derived from teasing? *Pan Africa News* 9:9–11.

INDEX

Page numbers in *italics* refer to illustrations.

acacia beans, 61
"acoustic universals," 166
acrobats, 94
adoptions, 24–26, *25*, 59, 132, 234–37
adrenaline, 267
affective computing, 277
affective neuroscience, 255–60
Affective Neuroscience: The Foundations of Human and Animal Emotions (Panksepp), 256
affiliation partners, 48
Africa, 40–41, 75–76
African grey parrots, 214–15, *226*, 227–28
afterlife, 44–45
Against Empathy (Bloom), 110–11
aggression, 33, 52, 74–76, 130, 134–35, 165, 183, 181–98, *193*, 259, 265, 291, 297, 300, 304
agricultural industry, 272
Agricultural Revolution, 191–92
"airplane game," 168
alarm calls, 163
Alex (parrot), 214–15
alpha females, 30–38, 43, 61–62, 97–98, 152, 164–65, 188, 197–202

alpha males, 28–30, 75–76, 77, 133, 143–44, 151–52, 164–65, 171–76, *174*, *178*, 180–88, 197–98, 201, 223–24
altruism, 98–99, 103–5, 114–20, 126
Alzheimer's disease, 119
American Sign Language, 162
Amos (chimpanzee), 40
amygdala, 50, 126, 194, 236
Anatomy Lesson, The (Rembrandt), 66
anesthesia, 269
anger, 65, 111, 166, 168, 169, 190, 232, 257, 258
animals:
  domestication of, 49, 83, 143, 229, 270–71
  habitats of, 69, 159, 246, 270–72
  hair raising by, 2, 21–22, 29, 33, 57, 63, 86, 141–44, 174, *174*, 178, *178*, 190, 232–33
  hierarchy of, 26–30, 31, 37–46, 61–63, 134, 142, 143, 145–46, 147, 150, 151–52, 177, 181, 186, 188, 200, 207, 223–24, 232, 276, 293
  human vs., 49–50

animals (*continued*)
  instincts of, 16, 86, 166, 189, 191, 196,
    204, 206, 276
  intelligence of, 48, 55–56, 79–80, 103,
    207, 232–33, 261, 269, 276–77
  laboratory, 107–8, 113, *117*, 215, 242,
    271–72
  offspring of, 2, 10, 20, 24, 40, 97, 103,
    104, 112, 130, 132, 148, 160, 163,
    173, 189, 194, 200, 228, 232, 236
  as pets, 41, 59, 89–91, 93, 96, 131, *145*,
    269, 270
  popular attitudes toward, 90–91,
    269–74
  power and status in, 27–28, 38–40,
    129–30, 141–42, 171–77, *179*,
    197–202
  as primates, 16–19, 32, 40–41, 44–45,
    52, *59*, 61, 70, 72–73, 76–77, 92, 95,
    99–100, 101, 106–7, 109, 113–14,
    131, 137, 138, 143–44, 146, 148,
    151, 153, *179*, 181, 190–97, 208,
    225, 272, 273–74
  rehabilitation centers for, 234–35, 271,
    272
  rights of, *91*, 269–74
  social behavior of, 7, 46, 99–100,
    108–9, 118–20, 129–35, 142, 154,
    162, 194, 209, 230–31, 233, 265,
    265, 267, 273–74
  species of, 52–53, 79–80, 168–69, 195,
    208, 248–49
  survival of, 98, 105, 118–19, 126–27,
    165–66, 208–9, 223, 231, 246, 248,
    254–55, 270
  zoos for, 1, 13–15, 17, 18, 20, 21, 23,
    24, 30, 36, 39, 43, 52, 56, 75, 112,
    113, 116–17, 121, 132–33, 143–44,
    158, 178, 181–82, 184, 186, 198,
    220, 242, 251, 270–72
  see also specific animals

Anonidium fruits, 197
anterior cingulate cortex, 222
anthropocentrism, 150–51
"anthropodenial," 50
anthropods, 249
anthropogenesis, 45
anthropoids, 51
anthropology, 195
anthropomorphism, 47–51, 55
antisocial behavior, 189–90, 202
antithesis principle, *144*, 174
anxiety, 63–64, 105–6, 127–30, 139, 183,
    188, 234–35, 258, 266–68, 272–73
ape "refugee" crisis, 271
apes, 24, 30, 43, 51–63, 66, 70, 76–80,
    86–87, 101, 103, 112, 114, 121, 130,
    132, 137, 148, 156, 168, 178, 192–95,
    208, 212, 217–28, *226*, 239, 249,
    270–71
apologies, 153–54
appendix, 166
aquariums, 264–65
archaeology, 193
Ardi (homonin fossil), 195, 197
Ardipithecus ramidus, *193*, 195
*Are We Smart Enough to Know How Smart
    Animals Are?* (de Waal), 16
Aristotle, 48, 75, 166, 176
Armstrong, Louis, 64
Arnhem colony, 20, 152, 197; *see also*
    Burgers Zoo
artificial intelligence, 276–77
associative learning, 159–60, 252, 255, 260
Astonishing Hypothesis, The (Crick), 222
athletics, 141–42, 145
Atlanta, 92, 95
Augustine, Saint, 224
Australopithecine, 44
Autism Spectrum Disorder, 96
automated responses, 255
axons, 266

baboons, 59, 61, 187–89
backslapping, 17, 80
bacteria, 166
Balinese temples, 138
Ballmer, Steve, 180
bananas, 217–18
barking, 29, 113, 122, 165, 189–90
Barrett, Lisa Feldman, 125, 256, 258
Bartal, Inbal Ben-Ami, 117–18
barter, 137–38, *137*, 210
Basic Emotion Theory (BET), 166
Baumeister, Roy, 231
beagles, 90–91, *91*
behavior:
  aggressive, 33, 52, 74–76, 130, 134–35,
    165, 183, 181–98, *193*, 259, 265,
    291, 297, 300, 304
  antisocial, 189–90, 202
  bonding (pair-bonding), 41–42,
    105–6, 141, 168, 182
  cooperative, 28, 99, 104, 110–20, 135,
    147, 164, 184, 197, 209, 214, 216–18
  cruel, 103, 107–9, 115–16, 208, 244
  cultural factors in, 53–54, 146, 158,
    167, 191, 195–97, 256, 258
  dominant vs. submissive, 4, 20–30,
    60, 62–64, 76, 80, 90, 97, 133, 135,
    141–46, *145*, 150–53, *174*, 177, 195,
    197–202, 223–25, 228–29, 276
  environmental factors in, 85–86,
    138–39, 247–48, 255, 270, 273–74
  evolution of, 44–45, 51, 52, 60, 68, 86,
    98–99, 106, 118, 132, 144, 147, 164,
    166–67, 170, 173, 194–97, 208–9,
    231, 243–44, 255–64, 270–71, 277
  experiments in, 80–82, 87, 113–14,
    116, 117–18, 136–37, 151–52, 159–
    60, 164, 165–66, 208–14, 233–34,
    250–51, 268–69, 286*n*
  fight-or-flight response in, 86, 232,
    276

  in fights, *3*, 77, 86, 109, 133, 135, 164–
    65, 172–76, 181–89, 232, 235, 276
  free will in, 35, 183, 221–31, 234, 270
  functional interpretation of, 126–27,
    167–68
  future orientation (expectation) in,
    42–43, 49–50, 136–39, 223–25, 250
  genetic background of, 45–46, 95,
    98–99, 168–69, 191, 192, 200, 222,
    241, 270
  grooming, 2, *3*, 14, 15, 19, 24, 28, 34,
    35, 40, 44, 61, 76, 106, 112–13, 118,
    130–31, 134, 149, 161, 163, 188,
    190, 194, 273
  helping, 28, 110–20
  imitative (mimicry), 70, 89, 93, 95, 159
  leadership, 180, 187–89
  learning and, 159–60, 252, 255, 260
  mechanistic view of, 262–63, 274, 276
  mediation in, 26, 28, 30, 38
  modification of (behaviorism), 8, 79–80,
    119, 255, 256, 259–60, 262, 274
  monogamous, 41–42, 105–6
  motivation in, 116–18, 167, 275–76
  needs and, 65, 113–14, 116, 143, 170
  philosophy of, 48, 75, 166, 176, 205–6,
    218–19, 222, 241, 262, 266
  play, 52–53, 71–74, 107–8
  predatory, 22–23, 45, 48–49, 62, 70,
    75, 86–87, 162, 187, 243, 253, 260,
    266–67, 276
  prosocial tendencies in, 114–16
  range of, 189, 272, 275, 277
  of "rational actors," 4, 211–31
  reflexive, 61, 249, 266, 268–69
  retaliatory (revenge), 132–34, 135
  rule for, *145*, 149–52
  science of, 8, 79–80, 111, 119, 129,
    255, 256, 259–60, 262, 274
  self-interest in, 98–99, 105 113–20,
    212–18

behavior (*continued*)
  sexual, 31, 32–35, *32*, *35*, 63, 80–81,
    114, 118–19, 126, 129, *137*, 138,
    148–49, 151, 162, 167, 168, 194,
    198, 219, 223–24, 243, 276
  signals in, 60–64, 70–71, 77, 144, 181
  social, 7, 46, 99–100, 108–9, 118–20,
    129–35, 142, 154, 162, 194, 209,
    230–31, 233, 265, *265*, 267, 273–74
  stimulus-response, 16, 260
  threat signals in, 61, 77, 126–27, 163,
    189–90, 247, 276
  victorious (pride), 140–46, *140*, 167
  violent, 36, 74–76, 109, 181–89, 202
  vocalization in, 24–25, 30, 47–48,
    154–55, 207–8
behaviorism, 8, 79–80, 119, 255, 256,
  259–60, 262, 274
Bekoff, Marc, 189
Bernstein, Carl, 179
beta females, 199–200
*Better Angels of Our Nature: Why Violence
  Has Declined, The* (Pinker), 195–96
biases, 138–39, 211–12
Bible, 110, 111, 192, 206
biomedical research, 271–72
biophilia, 270
"bipedal swagger," 143, *178*
birds, 60, 70, 81, 118, 132, 148, 161, 168,
  222, 227–28, 253, 254, 266–67, 275
blended emotions, 76–77
blind attraction, 100
blindness, 54
Bloom, Paul, 110–11
bluebirds, 222
blushing, 55, 147
Bobby (dog), 43
bodily fusion, 94
body language, 1, 6, 53–54, 80–81, 82,
  92, 124, 171–74
body signatures, 125

body temperature, 86–87
Bolt, Usain, 140
bonding, 41–42, 105–6, 141, 168, 182
bonding rituals, 141, 182
bonobos, 18, 43, 52, 56, 63, 75, 102, 112,
  114–15, *137*, 148–49, 152–53, 161,
  194–202, *193*, 234–35, 237
Borie (chimpanzee), 56–57
Borneo, 270–71
Born to Run (Springsteen), 30–31
Botox, 89
bottle feeding, *25*, 132
bowerbirds, 161
brachiators, 194
brain, brains:
  cognition and, *see* cognition
  damage to, 204–5
  emotions and, 126–28, 135–36, 155,
    156, 169
  imaging of, 89, 99, 105
  neurons in, 50, 84, 94–95, 99, 104,
    105, 128, 135–36, 233–34, 240, *241*,
    244, 266, 267
  neuroscience (neurology) of, 50,
    105, 168–69, 196, 233–34, 246–49,
    251–52, 255–63
  physiology of, 94, 105–6, 110, 135–36,
    194–95, 222, 252, 257–58, 259
  "second," 84
  size of, 16, 44–45, 50, 83, 107, 232–33,
    239–42, 244, 245, 248, 254, 255,
    260, 277–78
  waves of, 233–34
Braithwaite, Victoria, 268
Brazil, 208
breastfeeding, 148
breeding, 31–35, 48, 80–81, 114–19,
  148–49, 223–24, 243, 271
Brosnan, Sarah, 208–10, 212, 216, 220
Budongo Forest, 92–93
bullfighting, 271

Bully (dog), 150–51, 152, 154
Burgers Zoo, *14*, 15, 20, 43, 46, 133,
    181–82, 186–87, 251
Burkett, James, 105–6

cactus, 61
cadavers, 66
cages, 273–74
"call me!" signals, 60
calves, 236–37
camels, 132
canine teeth, 164–65, 198
captive breeding programs, 271
captivity, 50, 57, 184, 194, 271
capuchin monkeys, 10, 17, 87–88, 164,
    207–10, *211*, 221
carnivores (meat eating animals), 48–49,
    242–49, 272
carrots, 216
Castiles, 30–31
cats, 49, 55, 58, *59*, 73, 74, 81, 89–90,
    131, 143, 150, 156, 158, 223, 229,
    249, 259
caudate nucleus, 50
Ceausescu, Nicolae, 236
central nervous system, 249
cerebral cortex, 232–33, 256, 266, 278
charging, 21, 33, 74–75, 185, 186
chickens, 107–8, 274
children, 54, 70, 82, 93, 96–103, 115–16,
    119, 129, 177, *179*, 213, 217–18,
    222, 225–28, *226*, 229, 235, 264, 270
*Chimpanzee Politics: Power and Sex Among
    Apes* (de Waal), 175
chimpanzees:
    age of, 36–37
    as alpha females, 30–38, 43, 61–62,
        97–98, 152, 164–65, 188, 197–202
    as alpha males, 28–30, 75–76, 77, 133,
        143–44, 151–52, 164–65, 171–76, *174*,
        178, 180–88, 197–98, 201, 223–24

barking by, 29, 113, 122, 165, 189–90
breeding of, 31–35, 48, 80–81, 114–19,
    148–49, 223–24, 243, 271
captive, 50, 57, 184, 194, 271
charging by, 21, 33, 74–75, 185, 186
colonies of, 20–26, 77, 112–13,
    116–17, 121–22, 152, 175–76,
    185–86, 188, 197, 220–21, 234–35,
    251, 273–74
death of, 7–8, 13–17, 21, 23, 39–46,
    107, 109, 182–87
feeding habits of, 25–26, 60–61, 107–
    8, 114–15, 116, 130, 133, 159–61,
    192, 197–99, 208–12, *211*, 215–16,
    219–20, 250–53, 271
female, 20–21, 23, *32*, 70–71, 77,
    80–81, 112–13, 129–30, 138, 148,
    162, 164–65, 173, 182–83, 215,
    223–24
fieldworkers for, 49, 75–76, 199
grins, 3, 7, 13, *14*, 53, 60–64, 77, 87,
    165
grunting by, 13, 15, 36–37, 38, 57, 62,
    77, 108, 243
infant, 22, 40–41, 73, 97–98, 116–17,
    159–60, 163, 195–96, 228, 268–69
juvenile, 43 47, 52, 70, 71, 72, 92–93,
    97, 107–8, 130, *137*, 146, 164–65,
    178–79, 180, 188, 194, 215, 223,
    232, 271
male, *3*, 21 35, 80–81, 109, 129, 138,
    148, 172–74, *178*, *179*, 182–83, 243,
    276
nests of, 39, 40, 163, 182, 229–30, 253
night cages for, 39, 40, 182
peanuts eaten by, 32, 138, 155,
    160–61, 210, 253
political organization of, 26–27, 30,
    136, 171–80, *179*, 185, 191, 197–98,
    201–2, 217
ranking of, 37–40

chimpanzees (*continued*)
  refuges for, 272
  rivalry of, *3*, 28–30, *35*, 74–75, 109–10,
    141–44, 148, *178*, 180–86, 232
  territories for, 60, 109, 186, 189–90,
    265, 267
  trainers of, 60
  videos of, 17, 30, 39, 113, 273, 286*n*
  wild, 22–23, 92–93, 138, 194, 252–53
ChimpHaven, 272
Chrysippus's Dog, 250–51, 255
Church Fathers, 114
Churchill, Winston S., 191, 192
Churchland, Patricia, 104
cichlids, 265–66
circumcision, 269
Cirque du Soleil, 94
civilization, 195–97
Clay, Zanna, 235
Clayton, Nicky, 253
Clever Dog Lab, 213–14, *213*
climbing frames, 273
Clinton, Bill, 174
Clinton, Hillary, 173–74, 202
clock metaphor, 262–63
clownfishes, 264
clown loaches, 264–65, *265*
coalitions, 182–84
Coan, Jim, 233–34
cognition:
  biases in, 138–39
  language and, 5–6, 54, 116–17,
    123–26, 256–58, 269, 278
  logic and reasoning in, 206, 250–51
  memory and, 134 139–40, 252–54, 268
  meta-, 253–54
  mind vs. body dichotomy in, 84–85,
    88–91, 94, 97, 204–7, 221–22, 257,
    263
  objective vs. subjective perception in,
    87, 106–7

personality and, 46, 175, 207
rationality in, 110–11, 176, 207,
  211–18, 250–51
revolution in, 260, 262
self-awareness (consciousness) and,
  120, 123, 142, 144, 239–40, *241*,
  250–55, 260, 261–64, 277
tasks of, 77–78, 83, 88–89, 215, 221,
  232–33, 250–51, 264, 269–70, 277
in theory-of-mind, 263–64
Cognition Building, 82–83, 115
cognitive bias, 138–39, 221
"cold feet," 86
colonies, 20–26, 77, 112–13, 116–17,
  121–22, 152, 175–76, 185–86, 188,
  197, 220–21, 234–35, 251, 273–74
color vision, 159–60
Columbus, Christopher, 196
Comey, James, 171
communication, 54, 71–72, 80, 83–84,
  116–17, 123–26, 167, 256–58, 266,
  269, 278
  see also language
"concealed copulation," 148
conciliation, 134–35, 154–55, 184,
  190–91
condors, 271
conflict, 74–76, 135, 181–89, *193*, 259
conformity, 146–47, 160
conjugal relations, 32, 34
consciousness, 120, 123, 142, 144, 239–
  40, *241*, 250–55, 260, 261–64, 277
consolation, 101–3
contagious emotions, 97, 102–3, 118
cooperation, 28, 99, 104, 110–20, 135,
  147, 164, 184, 197, 209, 214, 216–18
corporations, 177, 181
corticosterone, 106
cortisol, 267
corvids, 41, 253
Costa Rica, 208

courtship rituals, 223–24, 276
cows, 236–37
coyotes, 143, 270
Crick, Francis, 222
crocodiles, 187, 245
cruelty, 103, 107–9, 115–16, 208, 244
crying, 24, 257–58
cryogenics, 206
cucumbers, 210–13, *211*, 220, 221
cultural factors in, 53–54, 146, 158, 167,
    191, 195–97, 256, 258
curare, 269
Curry, 62–63
cycling, 155

dairy industry, 236–37
Daisy, 40
Damasio, Antonio, 204–5, 246–47
Darwin, Charles, 7, 18, 55, 56, 67, 68,
    74, 84–85, 144, 147, 170, *174*, 205,
    222, 277
deafness, 24–25, 54
death protocols, 39
"death roll," 187
decision making, 84–85, 166–67
defensive reflexes, 61
democracy, 181
Democratic Republic of the Congo, 194,
    198
*Demonic Males: Apes and the Origins
    of Human Violence* (Wrangham),
    192–93
dendrites, 240
"Denver, the Guilty Dog," 149
depression, 24, 42, 268
Descartes, René, 262, 266
de Waal, Erica van, 159–60
diarrhea, 163
diazepam, 268
Diego (cat), 131
digestive system, 84, 166, 252, 258

Dimberg, Ulf, 88–89
diseases, 162–63
disgust, 17–18, 65, 68, 121–23, *122*, 126,
    127, 155–64, 166, 168, 169, 253, 258
distress, 97–98, 111–12, 116–18, *117*,
    194, 234–35
DNA, 95, 168–69, 191, 192, 222
Do Fish Feel Pain? (Braithwaite), 268
dogs, 42–43, 48, 49, 58, 59, 62, 71, 74,
    83, 89–91, *91*, 93, 96, 101, 107,
    111–12, 121, 131, 137–38, 144,
    149–52, 154, 156, 158, 163, 169,
    189–90, 201, 213–14, 229, 249,
    250–51, 252, 266
dolphins, 41, 49, 93, 132, 139, 245, 254,
    269
domestication, 49, 83, 143, 229, 270–71
dominance, 4, 20–30, 60, 62–64, 76,
    80, 90, 97, 133, 135, 141–46, *145*,
    150–53, *174*, 177, 195, 197–202,
    223–25, 228–29, 276
donkeys, 58
dopamine, 267
drugs, *117*, 258–59
    see also specific drugs
dualism, 49–50
Duchenne de Boulogne, Guillaume-
    Benjamin-Amand, 66–67
Duchenne smile, 66–67, *67*
duikers, 243
Dunlop, Tessa, 236

eagles, 187
Easterlin, Richard, 209
Easterlin Paradox, 209–10
economic systems, 99, 119, 211–12, 221
"ego strength," 227
Einfühlung ("feeling into"), 94
Ekman, Paul, 51–54, 58, 84, 123–24,
    125
electric shock, 233

elephants, 37, 43, 45, 80, 94, 102–3, 112, 113, 132, 139, 141, 143, 146, 168, 239–42, *241*, 249
Elliott (patient), 204–5
Elwood, Robert, 249
emasculation, 109
emoticons, 65
emotions:
  anger as, 65, 111, 166, 168, 169, 190, 232, *257*, 258
  basic (primary), 65, 123–25, 165–66
  biases in, 138–39, 211–12
  biology of, 53, 98, 166, 196–97, 208, 262
  blended, 76–77
  body as source of, 165–70, 257–58, *257*
  brain and, 126–28, 135–36, 155, 156, 169
  classification of, 65, 123–25, 165–70
  communication and, 54, 71–72, 80, 83–84, 116–17, 123–26, 167, 256–58, 266, 269, 278; *see also* language
  conciliatory (forgiveness), 134–35, 154–55, 184, 190–91
  contagious, 97, 102–3, 118
  control of, 35, 136, 221–30, *226*, 232
  cooperative, 231–37, 239, 243–44
  definition of, 84
  disgust as, 17–18, 65, 68, 121–23, *122*, 126, 127, 155–64, 166, 168, 169, 253, 258
  expression of, 1–2, 7–18, 65, 68, 22–23, 51–77, 83–89, 121–25, *122*, 126, 127, 155–68, 169, 231, 257–58, 277
  fear as, 19, 48, 61, 62, 65, 76, 85–86, 97, 126, 154, 166, 167, 176–77, 190, 203, 233–34, 253, 258
  feelings vs., 87, *91*, 125–28, 146, 167, 213–32, *241*, 255–60, 264, 275–77

guilt as, 21, 123, *145*, 147–54, 166, 169
happiness as, 65, 131, 169, *257*, 258, 277
intelligence based on, 48, 55–56, 79–80, 103, 203–37, 261, 269, 276–77
negative, 97–98, 218
optimistic (hope), 136–39
organs of, 165–70
rationality vs., 110–11, 176, 206, 211–18, 250–51
sentience and, 239–40, 254–55, 266, 269–74
stressful (anxiety), 63–64, 105–6, 127–30, 139, 183, 188, 234–35, 258, 266–68, 272–73
suffering and, 109, 176, 236, 245, 274
terminology of, 123–26
timeline for, 130, 139–40
"empathetic concern," 100–101
empathy, 49–50, 79–120, *117*, 153, 176, 189, 195, 212, 235, *241*
  see also sympathy
endorphins, 69
Engels, Friedrich, 261
English language, 5–6
enteric system, 84
environmental factors, 85–86, 138–39, 247–48, 255, 270, 273–74
envy, 26–27 33, 148, 189, 212–13, 218–19
epidemiology, 220
episodic memory, 252
Equine FACS (Ekman's Facial Action Coding System), 58
Erice conference (2016), 255–56
ethanol, 268
ethograms, 57
ethology, 126–27, 275
Etosha National Park, 141
eugenics, 111
eukaryotic cells, 246
European Union (EU), 177
euthanasia, 39, 271–72

evolution, 44–45, 51, 52, 60, 68, 86, 98–99, 106, 118, 132, 144, 147, 164, 166–67, 170, 173, 194–97, 208–9, 231, 243–44, 255–64, 270–71, 277

evolutionary cognition, 262

expectation, 42–43, 49–50, 136–39, 223–25, 250

experimentation, 80–82, 87, 113–14, 116, 117–18, 136–37, 151–52, 159–60, 164, 165–66, 208–14, 233–34, 250–51, 268–69, 286*n*

Expression of the Emotions in Man and Animals, The (Darwin), 7, 55

extramarital affairs, 148

eye contact, 2, 28, 63

Facebook, 158, 203, 219

face recognition, 81–83

"Faces and Behinds" experiment, 81–82

Facial Action Coding System (FACS), 51–54, 58

facial expressions, 4, 6, 7, 22–23, 51–77, *58*, *59*, 83–89, 104, *122*, 123–27, 129, 145, 155–58, 168, 180, 277–78

facial muscles, 66, 88

fairness, 110, 134, 207, *211*, *213*, 218–20, 244

fake smile, 66–68, *67*

families, 72, 148

Farage, Nigel, 174

farm industry, 236–37, 242–43, 272

fear, 19, 48, 61, 62, 65, 76, 85–86, 97, 126, 154, 166, 167, 176–77, 190, 203, 233–34, 253, 258

feathers, 161, 162

feeding, 25–26, 60–61, 107–8, 114–15, 116, 130, 133, 159–61, 192, 197–99, 208–12, *211*, 215–16, 219–20, 250–53, 271

Feeling of What Happens, The (Damasio), 246–47

feelings, 87, *91*, 125–28, 146, 167, 213–32, *241*, 255–60, 264, 275–77

female chimpanzees, 20–21, 23, *32*, 70–71, 77, 80–81, 112–13, 129–30, 138, 148, 162, 164–65, 173, 182–83, 215, 223–24

fertility, 32–33, *32*

Fessler, Daniel, 146

fetal position, 146

fieldworkers, 49, 75–76, 199

fighting, *3*, 77, 86, 109, 133, 135, 164–65, 172–76, 181–89, 232, 235, 276

fight-or-flight response, 86, 232, 276

Final Days, The (Woodward and Bernstein), 179

first-order fairness, 214–15

fish, 60, 264–68, *265*, 269, 275

Flehmen response, 58, *59*

flexitarianism, 244

fMRI scans, 169

food, 60–61, 107–8, 114–15, 116, 130, 133, 159–61, 192, 197–99, 208–12, *211*, 215–16, 219–20, 250–53, 271

foraging, 60–61, 252–53

foreheads, 232–33

forgiveness, 134–35, 154–55, 184, 190–91

fossils, 44–45, 193, 195, 197, 208

Foudouko (chimpanzee), 186, 188

France, 258

Frank, Anne, 91

Frankfurt, Harry, 223, 224, 227

free will, 35, 183, 221–31, 234, 270

French Revolution, 212, 216, 220

Freud, Sigmund, 196, 229

friendship, 48, 135, 142–43

frowns, 55–57, 88

fruit, 250–51, 253, 271

"funny faces," 52–53

future orientation, 42–43, 49–50, 136–39, 223–25, 250

Gaboon vipers, 43
Galileo, 114
games, 122
Gandhi, Indira, 201
ganglia, 248
gaping, 156
gaurs, 187
gaze direction, 147
Gazzaniga, Michael, 123
geese, 41, 141, 143, *140*
Geisha (chimpanzee), 39–40
gene pools, 270
genetics, 45–46, 95, 98–99, 168–69, 191,
    192, 200, 222, 241, 270
genital swelling, 32–33, *32*, *35*, 138
genocide, 111
George VI, King of England, 95
gibbons, 193–94
Gingrich, Newt, 175
Goblin (chimpanzee), 184–85, 188
goldfish, 266–67
Gombe National Park, 162–63, 169,
    184–85
Goodall, Jane, 20, 131, 162–63, 184–85
Good Samaritan, Parable of the, 110
Google, 180
goose bumps, 17
gorillas, 23, 49, 73, 190, 271
Grandin, Temple, 189, 256
grapefruit, *137*, 251
grapes, 210–13, *211*, 216, 220
Grasso, Richard, 212, 215
gratification, 223–28
gratitude, 130–32, 135
greasers, 30–31
great apes, 51, 63, *193*, 271
"great chain of being" (*scala naturae*), 66,
    245
greed, 99, 211–14
Greek civilization, 40, 48, 67–68, 205,
    250

Greg (elephant), 141, 143
greylag geese, 141
grief, 39–46, 92
Griffin (parrot), 214–15, *226*, 227–28
grimaces, 52–53
grins, 3, 7, 13, *14*, 53, 60–64, 77, 87, 165
grooming, 2, *3*, 14, 15, 19, 24, 28, 34,
    35, 40, 44, 61, 76, 106, 112–13, 118,
    130–31, 134, 149, 161, 163, 188,
    190, 194, 273
grunting, 13, 15, 36–37, 38, 57, 62, 77,
    108, 243
Gua (chimpanzee), 154–55, 229
guilt, 21, 123, *145*, 147–54, 166, 169

habitats, 69, 159, 246, 270–72
Hachiko (dog), 43
Hadza tribes, 192
Haeckel, Ernst, 222–23
hair raising, 2, 21–22, 29, 33, 57, 63, 86,
    141–44, 174, *174*, 178, *178*, 190,
    232–33
handholding, 233–34
hands, 60, 88
handshakes, 88
happiness, 65, 131, 169, *257*, 258, 277
Hare, Brian, 114
Hebb, Donald, 19, 26
heliotropism, 247
helping behavior, 28, 110–20
Henrich, Joseph, 212
Henry IV, King of England, 188–89
herbivores, 243, 244
herd instinct, 191
hermit crabs, 249
herons, 266–67
Hess, Walter, 259
hierarchies, 26–30, 31, 37–46, 61–63,
    134, 142, 143, 145–46, 147, 150,
    151–52, 177, 181, 186, 188, 200,
    207, 223–24, 232, 276, 293

high-wire artists, 94
Hobbes, Thomas, 72, 178, 196
Hoffman, Martin, 117
holistic observation, 2, 83
Holocaust, 91, 111
homeostasis, 246, 255
Homer, 48
hominids, 18, 76–77, 148, *193*, 195
hominization, 45
Homo economicus, 211–12, 221
homology, 106
Homo naledi, 44–45
Homo sapiens, 208
Hooff, Antoon van, 20
Hooff, Jan van, 13–21, *14*, 46, 51, 53, 59–60, 70, 76
hooting, 9, 83
hope, 136–39
hormones, 84, 104–6, 188
Horner, Vicky, 115
Horowitz, Alexandra, 149–50
horses, 48, 58, *59*, 93, 96, 128–29, 158
hospitals, 10, 39
How Animals Grieve (King), 40–41
Hrdy, Sarah, 104
Human: The Science Behind What Makes Us Unique (Gazzaniga), 123
Humane Research Council (HRC), 244
human-pig chimeras, 249
humans:
    ancestors of, 44–45
    animals vs., 49–50
    children of, 54, 70, 82, 93, 96–103, 115–16, 119, 129, 177, *179*, 213, 217–18, 222, 225–28, *226*, 229, 235, 264, 270
    cultures of, 53–54, 146, 158, 167, 191, 195–97, 256, 258
    economic systems of, 99, 119, 211–12, 221
    exceptionalism of, 50, 261

facial expressions of, 51–54, 88
girls vs. boys and, 97, 101
laughter in, 68–70
men vs. women and, 96, 110, 175, 201–2, 205
as parents, 10, 64, 72, 92, 100–101, 103, 146–47, 154, *179*, 230
warfare by, 189–97
wealth distribution and, 209–20, 221
Hume, David, 212–13
humor, 68–76, 122
humpback whales, 112
hunter-gatherers, 192, 196, 216, 220, 243–44
hunting, 195, 203, 216–17, 219, 234–36
hunting dogs, 250–51
huskies, 93
Hussein, Saddam, 171
hydrophobia, 116–17
hyenas, 45
hygiene, 161–63
hypothalamus, 259
hyraxes, 108

id, 229
identification, 91–92
Ig Nobel award, 82
imaging, 89, 99, 105
Imanishi, Kinji, 107
imitation, 70, 89, 93, 95, 159
immigration, 177
immune system, 156
impulse control, 35–36
incest, 162
income disparities, 209–20, 221
"inequity aversion," 212
infants, 22, 40–41, 73, 97–98, 116–17, 159–60, 163, 195–96, 228, 268–69
inferential reasoning, 250–51
injuries, 49, 92
input-output systems, 262–63

insects, 249

instincts, 16, 86, 166, 189, 191, 196, 204, 206, 276

insular cortex (insula), 155, 169, 194

intelligence, 48, 55–56, 79–80, 103, 203–37, 261, 269, 276–77

intentionality, 58–59, 187

internal rewards, 119–20

internal states, 53, 126–28, 255, 277

Internet, 17, 67, 111, 149, 156, 262, 286n

iPods, 96

irrationality, 211–18

isotocin, 267

"it takes a village" theory, 104

jackdaws, 41, 161

James, William, 227, 257–58, 260

James I, King of England, 250

Janmaat, Karline, 252–53

Japan, 36, 157, 160–61, 271

jealousy, 26–27 33, 148, 189, 212–13, 218–19

Jenny (orangutang), 56

Jericho, 192

Jews, 91, 110, 111

Jimoh (chimpanzee), 164–65

Johnson, Lyndon B., 90–91, *91*

jokes, 68–69

Jokia (cow), 102–3

Joni, 71, 100

"joy jumps," 74

judo, 142, 145

justice, 110, 134, *211*, *213*, 218–20, 244

juvenile chimpanzees, 43 47, 52, 70, 71, 72, 92–93, 97, 107–8, 130, *137*, 146, 164–65, 178–79, 180, 188, 194, 215, 223, 232, 271

Kafka, Franz, 239

Kakowet (bonobo), 112

Kalind (bonobo), 75

Kame (bonobo), 200

Kano, Takayoshi, 199

Kathy (greaser girl), 30–31

Katie (chimpanzee), 121–22

keas, 70

Kellogg, Donald, 230

Kellogg, Winthrop and Luella, 154–55, 230

Kennedy, John F., 201

Kenya, 61, 188, 271

"killer ape" theory, 195

killer whales, 112

Kim Jong-il, 180

King, Barbara, 40–41, 189

King's Speech, The (film), 95

kinship bonds, 134

kissing, 29

Kissinger, Henry, 179, 180

Klaus (chimpanzee), 232

knuckle-walking, 92, 193–94

koala bears, 158

Köhler, Wolfgang, 131

Koko (gorilla), 49

Koshima, 160–61

Kropotkin, Pyotr, 209

Kuif (chimpanzee), 23–26, *25*, 39–40, 43–44, 132

Kyoto, University of, 164

laboratory animals, 107–8, 113, *117*, 215, 242, 271–72

lactation, 24

Ladygina-Kohts, Nadia, 71, 100

Lahr, John, 69

Lamalera, Indonesia, 216–17

Lance (monkey), 221

language, 5–6, 54, 116–17, 123–26, 256–58, 269, 278

Lassie (dog), 42–43

"laugh face," 22–23, 53, 87

laughter, 22–23, 47, 53, 57, 59–60, 68–76, *74*, 87, 124, *125*

La vache qui rit (the laughing cow), 139
"law of effect," 260
leadership, 180, 187–89
learning, 159–60, 252, 255, 260
LeDoux, Joseph, 126, 128
legal rights, 271
Leipzig Zoo, 20
leopards, 163, 253
Lincoln, Abraham, 10, 107
Lindenplatz, 245–46
Linnaeus, Carl, 51
lions, 89–90, 96
Lipps, Theodor, 94
lip-smacking, 60–61
livestock, 236–37, 242–43
lizards, 142
lobsters, 249
logic, 207, 250–51
Lola ya Bonobo Sanctuary, 194, 234
London Zoo, 17–18, 89–90, 121
long-tailed macaques, 151
Lorenz, Konrad, 41, 107, 150–51, 191,
    192–93, 196
love, 41–42, 84–85, 168
loyalty, 23, 43
Luit (chimpanzee), 152, 154, 181–84,
    185, 187
Luke (dog), 42–43

Maasai Mara, 223
macadamia nuts, 36
macaques, 59, 63, 95, 134, 136–37, 138,
    143–44, 151, 160, 162, 272–73
Machiavelli, Niccolò, 178
Macron, Emmanuel, 176
Mae Perm (cow), 102–3, 119
Magee, Barry, 249
Mahale Mountains National Park, 108, 185
male chimpanzees, *3*, 21 *35*, 80–81, 109,
    129, 138, 148, 172–74, *178*, *179*,
    182–83, 243, 276

Mama (chimpanzee), 13–16, *14*, 13–16,
    18, 20–24, 26–30, 33–39, 71, 152, 197
mamans (local women guardians),
    234–35
mammals, 41–42, 48–50, 74, 93, 97, 102,
    103–6, 112, 117, 123, 128, 134–35,
    165–70, 245, 247, 249, 254, 267
mammary glands, 165
Mango (monkey), 208
mangrove crabs, 143
"mansplaining," 5–6
marine life, 248–49, 264–69
marriage, 72, 84–85, 205
marshmallow test, 225–27, *226*
Marx, Karl, 261
Mary Tyler Moore Show, 125
masks, 22–23, 75
Maslow, Abraham, 143–44
Masson, Jeffrey Moussaieff, 189
masturbation, 149
materialism, 222–23
matriarchs, 37–40, 139
May (chimpanzee), 92, 95
mazes, 250
McCain, John, 201
McConnell, Patricia, 42
meadow voles, 42
meat, 32–33, 48–49, 219
meat-eating animals (carnivores), 48–49,
    242–49, 272
mechanistic approach, 262–63, 274, 276
Medan, 127–28
mediation, 26, 28, 30, 38
Meir, Golda, 201
menstruation, 31–32, 84
Merkel, Angela, 201
metacognition, 253–54
meta-communication, 71–72
mice, 57, 97, 98
micro-expression, 89
Microsoft, 180

migration, 118
Milton, John, 221
mimicry, 70, 89, 93, 95, 159
mind vs. body dichotomy, 84–85, 88–91, 94, 97, 204–7, 221–22, 257, 263
minicapitalists, 212
mirror neurons, 94–95
"missing link," 193–94
mitochondrial DNA, 192
moats, 116–17
modeling, 92
modification of behavior (behaviorism), 8, 79–80, 119, 255, 256, 259–60, 262, 274
Moniek (chimpanzee), 28, 34–35
monkeys, 60–63, 76–77, 85, 87–88, 98, 156, 159, 192, 243, 245, 269, 286*n*
monogamy, 41–42, 105–6
Montaigne, Michel de, 92
moral values, 110–11, 132, 147, 156, 158, 164–65, 169, 208, 209–13, 218–19, 242–49, 269–74
morphine, 268
mortality, 39–46
mother-child relationships, 24–26 40–41, 97–98, 103–5, 129–30, 159–60, 163, 168, 202, 228
motivation, 116–18, 167, 275–76
mountain gorillas, 271
movies, 54, 75
murder, 109, 181–89, 191
muscle strength, 80
mussels, 248
Mussolini, Benito, 187–88
MythBusters (TV show), 86

naked mole rats, 162
Napoleon I, Emperor of France, 191
National Aeronautics and Space Administration (NASA), 20
Native Americans, 196

natto (fermented soybeans), 157
Natua (dolphin), 254
natural selection, 99, 166–67, 208–9, 263–64
needs, 65, 113–14, 116, 143, 170
negative arousal, 86
negative emotions, 97–98, 218
nesting, 161–62, 222
nests, 39, 40, 163, 182, 229–30, 253
Netherlands, 148, 155, 258, 271
neuroimaging, 99
neurons, 50, 84, 94–95, 99, 104, 105, 128, 135–36, 233–34, 240, *241*, 244, 266, 267
neuropeptides, 41
neuroscience (neurology), 50, 105, 168–69, 196, 233–34, 246–49, 251–52, 255–63
neurotransmitters, 50, 128, 267
New World, 196
New Zealand, 271
Nietzsche, Friedrich, 178
night cages, 39, 40, 182
Nikkie (chimpanzee), 28–30, 33–34, 38, 133, 136
Nishida, Toshisada, 185–86
Nixon, Richard M., 179
Nkombo (chimpanzee), 108
Noble Savage, 192
nociception, 266
nonverbal cues, 65, 68
nose wrinkles, *122*, 155–58
Nowak, Martin, 99
Ntologi (chimpanzee), 185–86
nuclear families, 148
Nummenmaa, Lauri, *257*, 258
nursing, 41–42, 103, 104
nut cracking, 36, 208

Obama, Barack, 174, 201
objective perception, 87, 106–7

O'Connell, Caitlin, 141
offspring, 2, 10, 20, 24, 40, 97, 103, 104, 112, 130, 132, 148, 160, 163, 173, 189, 194, 200, 228, 232, 236
Old Testament, 192
Olympics, 142
omnivores, 243
On Aggression (Lorenz), 191
On Bullshit (Frankfurt), 224
Oortje (chimpanzee), 33–35, 43–44
opioids, 256
optimism, 136–39
Orange (monkey), 61–62
orangutans, 17–18, 49–50, 56, 70, 97, 127–28, 225, 270–71
orgasm, 63, 126, *137*
Orgasmusgesicht (orgasm face), 63, *137*
orphans, 234–37
Ortega y Gasset, José, 123, 142
"ostracizing humor," 72
Othello (Shakespeare), 218–19
Ottoman Imperial Harem, 200
Oxford Companion to Animal Behaviour, The, 65–66
oxytocin, 41–42, 46, 104–5, 106, 267
oysters, 248

pachyderms, 102–3
pain, *58*, 97, 100–101, 107, 110, 248, 249, 255, 257, 266, 268–69
pair-bonding, 41–42, 105–6, 141, 168, 182
paleontology, 44–45, 208
Palin, Sarah, 201
Panbanisha (bonobo), 215–16
pandas, 158
pandiculation, 95–96
Panksepp, Jaak, 73–74, 255–56, 258, 259, 262, 263, 264
panthers, 22–23, 75
panting, 47–48, 69, 70–71

Papua New Guinea, 54
Paradise Lost (Milton), 221
parakeets, 270
Paralympic Games, 142
parasites, 162
parents, 10, 64, 72, 92, 100–101, 103, 146–47, 154, *179*, 230
Parr, Lisa, 84, 86
parrots, 70, 81, 214–15, *226*, 227–28
"Part Played by Labour in the Transition from Ape to Man, The" (Engels), 261
Pascal, Blaise, 85
pathogens, 157
pattern recognition, 2, 27
Payne, Katy, 94
peanuts, 32, 138, 155, 160–61, 210, 253
peer groups, 146–47
pelicans, 93
pennaceous feathers, 161
Peony (chimpanzee), 112–13, 115
Pepperberg, Irene, 214, 227
perception, 80–81
personality, 46, 175, 207
pessimism, 138
pesticides, 246
pets, 41, 59, 89–91, 93, 96, 131, *145*, 269, 270
pharaohs, 44
philosophy, 48, 75, 166, 176, 205–6, 218–19, 222, 241, 262, 266
Phineas (chimpanzee), 188
phobias, 126
pigs, 139, 249, 274
pigtail macaques, 59
Pilate, Pontius, 156
pilo-erections, 2
Pinker, Steven, 195–96
Pinky (rat), 73
pituitary gland, 41
plants, 245–48, 249

Plato, 222
play behavior, 52–53, 71–74, 107–8
"play bow," 71
"play faces," 52–53, 108
pleasure, *58*, 108, 110, 111, 126, 229,
    248, 255, 257
Plecostomus, 264
Plexie (cat), 49
Plotnik, Josh, 102
Plutarch, 93
poachers, 234–36
police, 190
polio, 162–63
politics, 26–27, 30, 136, 171–80, *179*,
    185, 191, 197–98, 201–2, 217
Poole, Joyce, 113
population control, 271–72
porpoises, 271
postmortem loyalty, 43
power, 27–28, 38–40, 129–30, 141–42,
    171–77, *179*, 197–202
prairie voles, 42, 105–6
"pre-concern," 98
predators, 22–23, 45, 48–49, 62, 70, 75,
    86–87, 162, 187, 243, 253, 260,
    266–67, 276
prefrontal cortex, 128
Premack, David and Ann, 250–51
pride, 140–46, *140*, 167
Primate Research Institute (Kyoto Uni-
    versity), 21, 273–74
primates, 16–19, 32, 40–41, 44–45, 52,
    59, 61, 70, 72–73, 76–77, 92, 95,
    99–100, 101, 106–7, 109, 113–14,
    131, 137, 138, 143–44, 146, 148,
    151, 153, *179*, 181, 190–97, 208,
    225, 272, 273–74
primatology, 27, 32, 106–7, 138, 148
Prince, The (Machiavelli), 178
problem-solving, 250–51
Proctor, Darby, 217–18

prosocial tendencies, 114–16
protein, 243
Proust, Marcel, 129, 252, 254
Prozac, 118
Pruetz, Jill, 186
psyche (soul), 67–68
psychology, 67–68, 79–80, 118–19, 158,
    177, 196, 219, 229, 234, 260–64
Putin, Vladimir, 201

Radboud University Nijmegen, 79
"rain face," *122*, 155, 161
Rand, Ayn, 99
ranking, 37–40
"rational actors," 4, 211–31
rationality, 110–11, 176, 206, 211–18,
    250–51
rational maximizers, 216–17
rats, 50, 57–58, 73–74, *74*, 79, 86,
    117–18, *117*, 126, 127, 129, 159,
    162, 209, 222, 250, 252, 256, 259
Rawls, John, 218–19
Reagan, Ronald, 99, 173
reasoning, 206, 250–51
recall, 252–53
red colobus monkeys, 243
reducetarianism, 244
reflexes, 61, 249, 266, 268–69
refugees, 91, 271
refuges, 272
rehabilitation centers, 234–35, 271, 272
Rembrandt, 66
Remembrance of Things Past (Proust), 252
Reo (chimpanzee), 21
Republican Party, 172
retaliation (revenge), 132–34, 135
rewards, 83, 119–20, 136–37, 209–15,
    211, *213*, 252
rhesus monkeys, 61–63, 80, 97, 134, 144,
    162, 273–74
rhinos, 271

Rita (chimpanzee), 115
ritualization, 60
rivalry, *3*, 28–30, *35*, 74–75, 109–10,
   141–44, 148, *178*, 180–86, 232
robots, 277
rodents, 57–58, *58*, 73–74, 105–6, 250,
   275
rollovers, 96
Romania, 236
Romney, Mitt, 172
Roosje (chimpanzee), 25–26, *25*
Rousseau, Jean-Jacques, 192

sadness, 39–46, 92, 65, 127, *257*
Safina, Carl, 189
St. Nicholas, 65
Salk Institute, 249
San Diego Zoo, 52, 75, 112, 220–21
Sapolsky, Robert, 169
Sapporo, 157
Sara (chimpanzee), 113
Sarabian, Cécile, 160
Sarah (chimpanzee), 250–51
Sarkozy, Nicolas, 171–72
savages, 192, 196
scala naturae ("great chain of being"), 66,
   245
Schadenfreude, 110
Schenkel, Rudolf, 175
Schilder, Matthijs, 22–23
Schoeck, Helmut, 219
Schreckstoff (shock substance), 267
Schwarzenegger, Arnold, 176
sclera, 58, 147
scratching, 93
screaming, 2, 9, 23, 28–29, 33, 43–44,
   97–98, 101–2, 116–17, 122, 141,
   147, 155, 176, 235–36
seals, 112
Sea of Cortez, 271
secondary emotions, 166

"secondary sisterhood," 199
"second brain," 84
second-order fairness, 214–15, 218
self-awareness, 120, 123, 142, 144,
   239–40, *241*, 250–55, 260, 261–64,
   277
self-control, 35, 136, 221–30, *226*, 232
self-esteem, 31, 48, 143–44, 175
selfies, 67
self-interest, 98–99, 105 113–20,
   212–18
"selfish genes," 98–99
Sendler, Irena, 111
Senegal, 186
sentience, 239–40, 254–55, 266, 269–74
serotonin, 267
sexual bargaining, 34–35, *35*, 129
sexual reproduction, 31, 32–35, *32*, *35*,
   63, 80–81, 114, 118–19, 126, 129,
   137, 138, 148–49, 151, 162, 167,
   168, 194, 198, 219, 223–24, 243,
   276
Shakespeare, William, 188–89, 218–19
shame and guilt, 123, 145–55, 169, *257*
"sham-rage," 259
Shapiro, James, 246
sharing patterns, 130–31
sharks, 107, 245
Sheila (chimpanzee), 113
Sherman, Rachel, 215–16
shore crabs, 249
signals, behavioral, 60–64, 70–71, 77,
   144, 181
Singer, Isaac Bashevis, 221
Singer, Tania, 110
skin color, 147
Skinner, B. F., 8, 79, 256
slapstick movies, 75
slavery, 10
sled dogs, 93
smartphones, 82

smell, *59*, 80–81
smiles, 64–67, *67*, 88
Smith, Adam, 105, 106, 114–15,
    137–38
snakes, 43, 62, 85, 107, 113
social behavior, 7, 46, 99–100, 108–9,
    118–20, 129–35, 142, 154, 162, 194,
    209, 230–31, 233, 265, *265*, 267,
    273–74
social sciences, 50, 98
songbirds, 161
South Africa, 229
South Korea, 113
space program, 20
Spain, 271
species, 52–53, 79–80, 168–69, 195, 208,
    248–49
species classification, 52–53, 168–69
sperm whales, 240
Spicer, Sean, 171
Spinoza, Baruch, 222
spontaneous behavior, 57
Springsteen, Bruce, 30–31
squirrels, 108
stalking, 161
status, 27–28, 38–40, 129–30, 141–42,
    171–77, *179*, 197–202
sterilization, 111
sticklebacks, 265–66, 275
stimulus-response, 16, 270
stress, 63–64, 105–6, 127–30, 139, 183,
    188, 234–35, 258, 266–68, 272–73
subcortical area, 259
subjective perception, 87, 106–7
submission, 4, 20–30, 60, 62–64, 76,
    80, 90, 97, 133, 135, 141–46, *145*,
    150–53, *174*, 177, 195, 197–202,
    223–25, 228–29, 276
suffering, 109, 176, 236, 245, 274
suicide, 18
Sumatra, 127–28, 225

*SuperCooperators: Altruism, Evolution, and
    Why We Need Each Other to Succeed*
    (Nowak), 99
superego, 229
survival, 98, 105, 118–19, 126–27,
    165–66, 208–9, 223, 231, 246, 248,
    254–55, 270
survival circuits, 126–27
survival niches, 270
"survival of the fittest," 208–9
Swaggart, Jimmy, 153–54
sympathetic locomotion, 93
sympathy, 79–80, 91–93, *91*, 100, 106–7,
    110–15, 118–19, 176
synchronization, 93
systematics, 168–69

taboos, 148, 151, 177, 237, 262
Taï National Park, 219, 252–53
*Take Pride: Why the Deadliest Sin Holds
    the Secret to Human Success* (Tracy),
    141–42
"Tantalus" game, 107–8
tantrums, 63, 102, 176–80, *179*, 219,
    232
Tanzania, 20, 192
Tara (chimpanzee), 122–23, 159
tarantulas, 86
taste aversion, 252
taxonomy, 248–49
Tchimpounga Rehabilitation Center,
    131
tears, 147
tension, 74–76
territories, 60, 109, 186, 189–90, 265,
    267
Tertullian, 109
Thailand, 102
thale cress, 247
thanatology, 40–41
Thanatos, 40

Thatcher, Margaret, 99, 201
Theory of Justice, A (Rawls), 218–19
theory-of-mind, 263–64
Theory of Moral Sentiments, The (Smith),
    105
Thomas, Elizabeth Marshall, 189
Thomas, Gospel of, 206
thought experiments, 250
threat signals, 61, 77, 126–27, 163,
    189–90, 247, 276
Through a Window (Goodall), 184–85
tickling, 47, 51, 73–74, 74, 256
tigers, 187
timeline, 130, 139–40
"time travel," 225
Tinbergen, Niko, 53
Tinka (chimpanzee), 93
Tinklepaugh, Otto, 136–37
tit-for-tat behavior, 133, 137–38
toilet training, 229–30
tokens, 217
Tokuyama, Nahoko, 199
Tolstoy, Leo, 89–90, 96
tones, 254
tools, 12
touchscreens, 81–83, 207
Tracy, Jessica, 141–42, 144, 145
trainers, 60
transparency, 272–74
trauma, 234–37
Treculia fruits, 197
trees, 245–46
triadic awareness, 29–30
Triumfgeschrei ("triumph-screaming"),
    141
Trudeau, Justin, 201
Trump, Donald, 172–76, 202
Trump, Eric, 175
trust, 15, 102, 147, 175, 220
Tulp, Nikolaas, 66
Twain, Mark, 55, 131

Ultimate Fighting Championship, 64
Ultimatum Game, 215–18, 225
ungulates, 10

vaquitas, 271
vegan, 244–45
vegetarianism, 244–45
ventromedial frontal lobe, 204–5
Venus flytraps, 247
verbal communication, 54, 80, 116–17,
    123–26, 256–58, 269, 278
    see also language
Vernon (bonobo), 75
vertebrates, 165, 256
vervet monkeys, 159–60, 228–29
"vibes," 88
Victoria, Queen of England, 17–18, 121
victorious behavior (pride), 140–46, *140*,
    167
videos, 17, 30, 39, 113, 273, 286n
Vietnam War, 90, 191
Vilas Park Zoo, 143–44
violence, 36, 74–76, 109, 181–89, 202
    see also warfare
"visceral disgust," 156
visceral language, 257–58
vision, 80–81
vocalization, 24–25, 30, 47–48, 154–55,
    207–8
voles, 42, 105–6
vomeronasal organ, *59*
voodoo dolls, 134

Wall Street, 99
wantons, 221–23
warfare, 189–97
Warneken, Felix, 119
Warsaw ghetto, 111
Washoe (chimpanzee), 116–17, 119
"water need," 128
wealth distribution, 209–20, 221

Wealth of Nations, The (Smith), 105
Wedgwood, Emma, 84–85
Westermarck, Edvard, 132
western scrub jays, 253
whales, 112, 216–17, 240
When Elephants Weep: The Emotional Lives of Animals (Masson), 189
whole-body image, 82
whooping calls, 225
wild chimpanzees, 22–23, 92–93, 138, 194, 252–53
wildebeests, 223
Williams, Serena and Venus, 125
Wilson, E. O., 270
winner-take-all mentality, 220
Wittgenstein, Ludwig, 224
wolves, 57, 70, 139, *174*, 175, 209, 214
woodpeckers, 60
Woodward, Bob, 179
World War II, 111, 191
Wounda (chimpanzee), 131
Wrangham, Richard, 192–93
wrestling, 72

Xenophanes, 48
xenophobia, 194

Yamamoto, Shinya, 116
yawning, 95–96
Yerkes, Robert, 32–33
Yerkes National Primate Research Center, 1–2, 19, 40, 56–57, 77, 82–83, 121, 164–65, 207, 273–74
Yeroen (chimpanzee), 29–30, 38, 152, 186–87
Young, Larry, 42
Yucatec Maya, 126

Zamma, Koichiro, 108
zebrafish, 268
zebras, 58, 203
Zoological Society of London, 56
zoon politikon ("political animal"), 176
zoos, 1, 13–15, 17, 18, 20, 21, 23, 24, 30, 36, 39, 43, 52, 56, 75, 112, 113, 116–17, 121, 132–33, 143–44, 158, 178, 181–82, 184, 186, 198, 220, 242, 251, 270–72

ABOUT THE AUTHOR

Frans de Waal is a Dutch-American ethologist and zoologist. Having earned a Ph.D. in biology from the University of Utrecht in 1977, he completed a six-year study of the chimpanzee colony at Burgers Zoo, in Arnhem, before moving to the United States. His first popular book, *Chimpanzee Politics* (1982), compared the schmoozing and scheming of chimpanzees involved in power struggles with that of human politicians. Ever since, de Waal has drawn parallels between primate and human behavior. Translated into over twenty languages, his books have made him one of the world's most visible biologists.

With his discovery of reconciliation in primates, de Waal pioneered research on animal conflict resolution. He received the 1989 *Los Angeles Times* Book Award for *Peacemaking Among Primates*. His scientific articles have been published in journals ranging from *Science*, *Nature*, and *Scientific American* to those specializing in animal behavior and cognition. His latest interests are animal cooperation, the origin of empathy, and the evolution of human morality.

De Waal is C. H. Candler Professor in the Psychology Department of Emory University, Director of the Living Links Center at the Yerkes National Primate Research Center, in Atlanta, and Distinguished Professor at the University of Utrecht. He is a member of both the U.S. National Academy of Sciences and the Royal Netherlands Academy of Arts and Sciences. In 2007 *Time* voted him one of the World's 100 Most Influential People Today.

With his wife, Catherine, he lives in Smoke Rise, Georgia, USA.

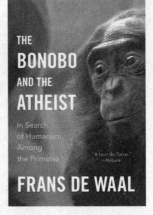